Changing Land Management

In memory of Amabel Fulton

Changing
Land Management

Adoption of New Practices by Rural Landholders

Editors: David Pannell and Frank Vanclay

CSIRO

PUBLISHING

National Library of Australia Cataloguing-in-Publication entry

Changing land management : adoption of new practices by rural landholders/edited by David Pannell and Frank Vanclay.

9780643100381 (pbk.)
9780643101739 (eBook)
9780643102279 (ePub)

Includes bibliographical references and index.

Farm management – Decision making.
Land use, Rural – Planning
Land use, Rural – Environmental aspects.
Agriculture – Economic aspects.
Farmers – Finance, Personal.

Pannell, David J., 1960–
Vanclay, F. M. (Francis M.)

630.685

Published by
CSIRO PUBLISHING
36 Gardiner Road, Clayton VIC 3168
Private Bag 10, Clayton South VIC 3169
Australia

Telephone: [+613] 9545 8555
Local call: 1300 788 000 (Australia only)
Fax: +61 3 9662 7555
Email: csiropublishing@csiro.au
Web site: www.publishing.csiro.au

Cover image by Eloise Seymour

Set in 10/12 Adobe Minion Pro and ITC Stone Sans
Cover and text design by James Kelly
Typeset by Desktop Concepts Pty Ltd, Melbourne
Printed by Ingram Lightning Source

CSIRO PUBLISHING publishes and distributes scientific, technical and health science books and journals from Australia to a worldwide audience and conducts these activities autonomously from the research activities of the Commonwealth Scientific and Industrial Research Organisation (CSIRO).

Feb26_RP_ILS

Contents

Contributors

Neil Barr is a social researcher with the Department of Primary Industries, Victoria.

Denise Bewsell is a researcher in the Social Research Team in AgResearch, Christchurch, New Zealand.

Ray Cooksey is Professor, Faculty of The Professions, School of Business, Economics and Public Policy at the University of New England.

Allan Curtis is Professor of Environmental Management at the Institute for Land, Water and Society and Head of Campus – Albury/Wodonga, at Charles Sturt University.

Amabel Fulton completed a PhD in extension at the University of Tasmania in 2009. Together with her husband, David Fulton, in 1999 she started the consulting company Rural Development Services, which seeks to enhance human and organisational capacity in rural and regional Australia. Amabel passed away in 2009 just as the chapter was being finalised.

Geoff Kaine is a social researcher with the Department of Primary Industries, Victoria.

Rick Llewellyn is a farming systems scientist with CSIRO Sustainable Ecosystems based in Adelaide, South Australia and a researcher in the Future Farm Industries Cooperative Research Centre.

Graham Marshall researches institutional-economic issues in environmental governance at the Institute for Rural Futures at the University of New England.

Cathy McGowan AO is the Managing Director of a consulting company and member of a network of independent skilled professionals who engage with communities around issues of leadership, rural development and empowerment.

Emily Mendham is post-doctoral fellow (social science) with the Science into Society Group, CSIRO Earth Science and Resource Engineering Division in Brisbane, Queensland. She completed her PhD at Charles Sturt University in 2010.

David Pannell is ARC Federation Fellow and Professor, School of Agricultural and Resource Economics, and Director of the Centre for Environmental Economics and Policy, University of Western Australia. He is a researcher in the Future Farm Industries Cooperative Research Centre.

Frank Vanclay is now professor in the Department of Cultural Geography at the University of Groningen in The Netherlands. He was previously professor of rural sociology at the Tasmanian Institute of Agricultural Research at the University of Tasmania.

Roger Wilkinson is a social researcher with the Department of Primary Industries, Victoria.

Vic Wright is Adjunct Associate Professor, School of Business Economics and Public Policy at the University of New England.

Preface

There is a rich and extensive history of research into factors that encourage farmers to change their land management practices, or inhibit them from doing so. Many clear and important lessons have been learnt, but surprisingly, these are often apparently unknown or ignored by organisations that attempt to influence the decisions of farmers and other rural landholders.

The origin of this book was a meeting of Australian social researchers held in May 2001 to discuss their priorities for work within a new national research centre, the Cooperative Research Centre for Plant-Based Management of Dryland Salinity. Three of the contributors to this book (Pannell, Curtis and Barr) were present at the meeting, which agreed that the priority for the new centre was not to conduct new research into adoption, but to review synthesise and communicate existing knowledge in a way that would be useful to managers and biological scientists in the centre. The team was expanded to include Vanclay, Marshall and Wilkinson, and proceeded over the next several years to prepare an interdisciplinary review paper, 'Understanding and promoting adoption of conservation practices by rural landholders', which ultimately appeared in the *Australian Journal of Experimental Agriculture* in October 2006.

As well as being useful to the research centre, we found that the paper was of great interest to many others. It rapidly rose to the top of the journal's list of most downloaded papers and has since been cited over 80 times.

Given that there was little new in the paper, we were surprised at this level of interest. Clearly there was an unmet demand for this information. To respond further to this demand and to celebrate the success of the review paper, we organised a national workshop in Melbourne in November 2008 (with a sequel in Perth in July 2009). Again, we were surprised by the level of interest, finding that we had to move the event to a much larger venue to accommodate the number of delegates who registered for the Melbourne meeting (over 400).

Given the continuing intense interest, the broadened team, now including Kaine and Llewellyn, agreed that we would capture the event in a variety of ways: as audio recordings, video recordings and PowerPoint files, all of which were made available on the internet after the event (www.ruralpracticechange.org). In addition, we agreed to prepare a book, to allow each of the team to expand on their particular areas of interest and to present findings and insights from recent research.

Once again, the team was expanded to cover particular issues in more depth (Fulton and McGowan) and various co-authors (Bewsell, Cooksey, Mendham and Wright) were also involved. The result is this book.

Throughout, there are terms used that are common to this field of endeavor but which may be unfamiliar to some readers. In Chapter 1, we have provided a concise definition with the first instance of each such term.

We are grateful to many organisations and individuals for their support for the various elements of this campaign to improve knowledge about this research and its implications. The Cooperative Research Centre for Plant-Based Management of Dryland Salinity and its successor, the Future Farm Industries Cooperative Research Centre, provided the impetus and support in many ways. A number of funding bodies have supported the research of the contributors reported here, including Land and Water Australia, Rural Industries Research and Development Corporation, Grains Research and Development Corporation, the Australian Centre for International Agricultural Research, the Commonwealth Environmental Research Facilities (CERF) program, the Australian Water Research Advisory Council, and the Australian Research Council. Financial support for the workshops was provided by the Department of Primary Industries Victoria, Landmark, Future Farm Industries Cooperative Research Centre, the University of Melbourne, the University of Western Australia, Rural Industries Research and Development Corporation, the Australian Agricultural and Resource Economics Society, the Department of Agriculture and Food Western Australia, South Coast Natural Resource Management, Avon Catchment Council and Northern Agricultural Catchments Council. BoaB Interactive handled multimedia recording and website development. Other contributors include Fay Davidson, Chris Davies (from Tweak Editing), Kevin Goss, Andrew Campbell, Anna Roberts, Rowan Reid, Lauren Rickards, Sze Flett, Romy Greiner, Sally Marsh, Frank D'Emden, Karen Barroga, Perry Dolling, Peter Sullivan, Justin Bellanger, Rolan Deutekom, Leigh Sparrow, Aysha Fleming, Ingrid van Putten, Sally Marsh, Geoff Kuehne and Michael Robertson. We are grateful to all of these contributors.

David Pannell and Frank Vanclay
October 2010

Changing land management: multiple perspectives on a multifaceted issue

David J Pannell and Frank M Vanclay

Summary

Although the adoption of new practices by land managers has been well studied over many decades, much of that accumulated knowledge and understanding is not appreciated by most policy makers, extension officers, agricultural consultants, agricultural research scientists, and their managers. In this book, multiple perspectives on this multifaceted issue are identified, and a coherent understanding that combines these perspectives is presented. Common themes are the need for an appreciation of the diversity of land managers and their contexts, and for this diversity to be reflected in the way policy, governance arrangements and extension programs are designed. A more sophisticated understanding of the adoption process would improve the design and implementation of policy, governance and extension, and would lead to the greater achievement of desired outcomes.

Introduction

Many entities have an interest in promoting changes in rural land management, including governments, regional natural resource management agencies, other environmental management agencies, and non-government organisations such as Greening Australia, farmer organisations and Landcare groups. Their interests in encouraging change occur sometimes because of increased awareness of the environmental harm resulting from current land management practices, and sometimes from a desire to increase the profitability of rural industries and the well-being of rural people. At different times over past decades, phenomena such as soil erosion, salinity and other forms of land degradation, and more recently, climate change and biosecurity issues, have rallied governments and communities to action (Vanclay and Lawrence 1995). Effective facilitation of change in land management requires an understanding of the complex social, psychological and economic dynamics that affect decision making by land managers. This book is intended to expound and explain that understanding for the benefit of policy makers, extension officers, agricultural consultants, agricultural research scientists, and their managers and supervisors.

The adoption of new practices by land managers has been studied by many researchers over the past 50 years or so (e.g. see reviews by Feder *et al.* 1985; Feder and Umali 1993; Knowler and Bradshaw 2006; Marra *et al.* 2003; Pannell *et al.* 2006; Vanclay 2004; Vanclay and Lawrence

1995), and from a number of different disciplinary perspectives – e.g. rural sociology, economics, social psychology, agricultural extension – all of which are represented in this book. The phenomenon of practice change by rural landholders is generally very well understood, at least by those researching the topic. However, practitioners are not always aware of this literature, and do not necessarily understand all of the drivers (and their interactions) that influence the decision making of land managers.

In presenting a range of perspectives about the factors that affect land managers and their decision-making processes, and in providing advice about how to enable practice change in land management effectively, this book will be useful to different groups of readers, including:

- Policy makers who need to understand and anticipate the likely response of land managers to new policies and governance arrangements. For example, the success of environmental policies that aim to change land management will depend on the responsiveness of land managers to the specific policy mechanisms used. Their responsiveness will vary for different issues, in different regions, and for different policy mechanisms.
- Agricultural scientists and science managers who need to understand whether farmers and other landholders are likely to adopt recommendations resulting from their research, and if so, over what scale. Such understanding can help scientists target their research efforts in a way that is most likely to generate benefits.
- Environmental managers who need to understand the likely responses of different types of land managers to environmental projects. Will they participate, and if so, at what scale? What would it take to convince landholders to take part in a project? Which delivery mechanisms are most appropriate for encouraging changes in land management in a particular project?
- Extension agents and agricultural consultants who need to understand what motivates their clients, in order to design and deliver their extension programs. They also need to understand the limits of extension, so that resources are not wasted promoting changes that are of no interest to land managers.

These groups and organisations are interested in the results of research into practice change by rural landholders. What is surprising, however, is that the knowledge and understandings presented in this book are not already widely understood and embedded in their thinking and practice. Indeed, we commonly observe organisations behaving in ways that reveal their lack of knowledge of the research, and/or perhaps their unwillingness to accept it. Policies and governance arrangements are often designed based on naïve and inaccurate assumptions about landholder behaviour; agricultural scientists often develop new technologies or practices and are then surprised and disappointed at the limited extent to which they are adopted; environmental managers often express frustration at landholders' limited response to environmental programs; and extension agencies often apply information-based extension strategies in situations where they are most unlikely to result in changes in landholder behaviour.

As scholars and practitioners of the science of behaviour change, we should not be surprised at the above observations. We know from our research and experience that it is often difficult to get new ideas adopted. We know that their adoption depends on a whole host of factors. Throughout the chapters of this book, the many relevant factors are discussed from a variety of perspectives, in the context of land management.

The process of enabling change

Changes in land management practices happen as a result of a number of different processes. Individual land managers, usually in conjunction with the 'significant others' in their family farm businesses, may base their decisions on the results of their own experimentation, adaptation and invention. They may respond to market signals, or to information that they actively seek. Alternatively, they may be more passive and respond to information provided to them from a variety of sources, including private-sector agricultural consultants, field staff of agricultural processing companies and wholesalers, or extension staff from natural resource management agencies, environmental non-government organisations, or government departments. Some land managers are innovative in character and like trying different things, whereas others are less likely to try new approaches. Many of the authors in this book discuss the various stages on the adoption curve, highlighting the fact that there is more to the issue than identifying the innovators and laggards.

Organisations can encourage practice change in various ways. In the past, government agencies wishing to promote change in agriculture relied primarily on traditional extension approaches, involving the provision of advice to farmers – either one-on-one, in groups, or by distance using print media such as farm factsheets. While effective communication remains important, it is only one of the policy instruments used by organisations in the process of enabling change. Other policy instruments that are relevant in particular cases include covenants and memoranda of understanding, market-based mechanisms (such as conservation tenders and cap-and-trade systems), direct funding, regulation, and self-regulation through industry codes of practice. The effective promotion of change requires the right blend of policy instruments.

Recognising that information provision may not be sufficient, thinking within the field of extension has evolved. For example, the State Extension Leaders Network defines extension 'as the process of enabling change in individuals, communities and industries involved in the primary industries sector or with natural resource management' (SELN 2006, p. 2). They view the role of extension agents as building the capacity of people to make effective decisions and to deal with changing circumstances. Thus, rather than telling land managers what to do, extension agents would seek to ensure that land managers have the resources they need to make their own decisions in an informed way (SELN 2006).

Overview of chapters

The origins of this book lay with the review paper, 'Understanding and promoting adoption of conservation practices by rural landholders' (Pannell *et al.* 2006), which was prepared by six of the authors published here. The considerable interest in this paper led to further collaboration and ultimately to this book (see the history given in the Preface). That paper, reproduced here as **Chapter 2**, provides a synthesis and interpretation of a large body of past research literature from around the world. Although the focus is on adoption of conservation practices, it is in fact relevant to adoption of new practices generally.

Adoption is a process of learning, in which landholders are initially uncertain about the effectiveness of a new practice, but this uncertainty wanes over time as they accumulate information and personal experience. During this learning process, landholders move through several phases of the adoption process: awareness, non-trial evaluation, trial evaluation, adoption, review and modification, and if they do adopt (non-adoption is a possible outcome) they may ultimately 'disadopt'. To disadopt means to discontinue use of a practice. Wilkinson discusses the process of disadoption in Chapter 3.

Social, cultural and personal influences on adoption decisions are discussed. The extent and speed of adoption of new practices depends on a variety of aspects of the links between landholders and others. For example, adoption is influenced by networks, physical proximity, relationships and the actions of government programs. Differences in adoption decisions have also been explained by differences in the characteristics of landholders themselves, including their degree of motivation for financial profits, age, education, and the sizes of their properties. The role of the family is discussed as a crucial influence for many decisions.

The other key drivers of adoption decisions are the characteristics of the new practices. These are discussed in two categories: factors that influence the 'relative advantage' of the practice (the extent to which the practice is 'adoptable', meaning that adoption would be superior to non-adoption given the goals and preferences of the landholders – its advantages overall outweigh its disadvantages) and factors that influence the 'trialability' of the practice. (Trialability means how easy it is to learn about the new practice from a small-scale trial. A practice that is trialable is easy to learn about. The concept is discussed in detail in Chapter 2.) There are many factors in both categories. Relative advantage is influenced, for example, by the innovation's profitability, effects on the farming system, adjustment costs, riskiness, compatibility with existing practices, complexity, and the influence of government policies. Influences on trialability include the ability to trial the practice on a small scale, the observability of trial results (the ease with which results can be observed or detected), the time-lag between undertaking a trial and being able to observe its results, the difficulty of interpreting results, and the cost of a trial, amongst other factors. Implications for researchers, for extension agents and for policy makers are outlined. These implications, as well as some other points, are explored in more depth in some of the later chapters, each of which examines a more focused subset of the issues related to practice change by rural landholders.

Chapters 3 and 4 provide additional 'big picture', broad-ranging views on adoption of new practices by rural landholders. In **Chapter 3**, Roger Wilkinson emphasises that the concept of 'adoption' of a new practice is complex and multifaceted. Adoption is not an event but a process, often involving changes over time in the nature or extent of use of the practice. Adoption does not necessarily reach a steady state; that is, for most innovations the process of reviewing the performance of the innovation and reconsidering its extent and types of use continues even when it is being used in a stable way and to a high level. Changes in market conditions, for example, may lead to changes to the extent of a given practice's use at any time. Some landholders may adopt the practice in its entirety, using it to the fullest possible extent, while others may adopt it only partially, using it in specific niches or only occasionally. Clearly the same technology may be used in different ways by different landholders. Some innovations consist of packages of related technologies. In these cases, landholders rarely adopt the package as a whole, especially in the first instance. Rather they trial particular elements of the package, mixing and matching to suit their own needs. While this is often frustrating to the researchers who developed the original package, it is a legitimate and very sensible approach for landholders to take. In a similar vein, even for innovations with only one element, landholders often adapt the technologies to suit their own needs and circumstances more closely. In a sense, the technology that is adopted may be different from farm to farm, making it difficult to compare adoption between farms, or to aggregate the level of adoption. Finally, adoption is not forever. A practice may be superseded by a superior technology, may become unattractive due to changes in markets, or may serve a purpose that becomes redundant, resulting in disadoption. A key message of the chapter is that people need to take care in their usage of the word 'adoption', being clear about exactly what it means in their case, and the extent to which the ambiguities outlined by Wilkinson are relevant.

In **Chapter 4**, Frank Vanclay presents a large number of 'social principles' related to the adoption of new practices. While his discussion is primarily addressed to extension agents working in agriculture, the general understanding is also very relevant to other groups with an interest in rural practice change. The key message of the chapter is that farming is not just a technical or economic activity, but is very much a socio-cultural practice, influenced by a range of social processes. These processes influence the adoption of new practices by farmers. For example, a social view of farming recognises that women are an integral part of the farm and that family considerations play an important role in determining what gets done on the farm and why. One aspect of the family circumstance is its stage in the family life cycle (e.g. families with young children, families with adolescents, etc.), which influences factors such as the need for disposable income and what finance might be available for other purposes (such as investing in the trialling of new practices). Vanclay emphasises that extension agents need to understand the worldviews of farmers, and that this is made more challenging by the fact that farmers are a highly heterogeneous group. Understanding the adoption process from the farmer's point of view gives insight into the reasons for non-adoption of particular practices, and leads to the view that there are generally legitimate reasons for that non-adoption. Vanclay argues that science and extension do not automatically have credibility and legitimacy with farmers – suggesting that these have to be earned, which can be a slow and somewhat fragile process. For this and other reasons, realistic expectations for the impact of extension on adoption of new practices by farmers should usually be modest. As outlined throughout the book, there are many factors that play a role in adoption, and extension cannot be expected to convince farmers to adopt a practice that they have judged is not in their interests.

Chapters 5 and 6 are both relevant to strategic thinking about adoption and how to target efforts to increase it. In **Chapter 5**, Geoff Kaine and colleagues present a method for identifying the population of potential adopters of an agricultural innovation. The process is based on the recognition that different farmers obtain different benefits from any given innovation, depending on their farming contexts. For example, different farms have different resources (land area, soil types, machinery, finance), different economic circumstances (debt, distance from market), and different farmers have different personal circumstances (number of children, attitude to risk, etc.). The process segments the population of farmers according to these different farming contexts, and studies the likely nature and extent of adoption within each segment. It starts with a qualitative phase, in which farmers are engaged in interviews designed to understand the relevant farm contexts driving adoption decisions for a particular innovation. This is followed by a quantitative phase, in which a broader sample of the farming population is surveyed, using questions based on insights from the first phase. The approach is illustrated with a real example: the adoption of micro-irrigation technologies in horticulture. The example shows how the approach can provide a rich and nuanced understanding of the prospects for adoption of a particular practice. It confirms Vanclay's argument that once the different segments of the population are identified and their contexts understood, expectations about the overall extent of adoption will be much more modest than would be derived from a more naïve assessment. On the other hand, this sort of insight may lead to a reassessment of the success of extension. An extension program that leads to adoption by a relatively small proportion of the total population may in fact be judged as successful in the sense that a high proportion of potential adopters are adopting.

In **Chapter 6**, Rick Llewellyn outlines an approach for targeting extension efforts. It consists of identifying perceptions that are influential in the adoption decision and likely to be influenced by the provision of information by extension agents. Many perceptions are potentially important in the decision to adopt an agricultural innovation. For example, adoption of a new herbicide may be influenced by views about its effectiveness in controlling weeds, the market price of grain, the effects of the herbicide on human health, and the type of tillage system preferred by the farmer. In addition, there are apparently relevant factors that may not be highly influential with regard to the adoption decision. Perhaps health or environmental considerations are not important to a particular population of farmers. An understanding of which factors are and are not influential can be used to guide extension agents when they choose which issues to focus on in their extension activities. However, the fact that farmer perceptions about a factor have the potential to influence adoption does not necessarily mean that they *will* be influential. In particular, if farmers' perceptions about that factor are already accurate, providing additional information about it will not alter their decision. Thus, efforts to provide information to farmers about a technology should focus on those factors that will influence the adoption decision and which are not currently perceived accurately. A factor that is perceived incorrectly by farmers may not necessarily be a good target for extension, if that factor is not an important influence on adoption. Conversely, a factor that is an important influence on adoption, may not be worth providing information about if it is already accurately perceived. The process described in the chapter provides a systematic approach to understanding these issues and responding appropriately.

Family farming remains the dominant form of agriculture in Australia, notwithstanding continuing economic pressures for farms to expand to remain viable. In **Chapter 7**, Amabel Fulton and Frank Vanclay focus on issues related to the farm family, emphasising that a good understanding of these issues is often important when designing mechanisms to encourage changes in farm practice. For example, such understanding may allow agricultural extension to engage with the right people at the right time to maximise the prospects for change. Extension conducted within the 'product push' paradigm may fail to recognise differences in needs, perspectives and preferences between different family members, such that it fails to deliver meaningful and highly valued change. The authors emphasise the value of a 'needs-based' approach to extension that is more responsive to the needs of individual farm businesses. This poses challenges to the public extension system, which has tended to move away from individualised extension over the past two decades. On the other hand, private farm business consultants tend to operate in a way that is much closer to the mode of operation that is advocated in this chapter.

As well as the family, the community within which a farmer lives and operates also influences the farmer's decisions about new practices. In **Chapter 8**, Graham Marshall outlines the findings and implications of a relatively recent body of research exploring the relationships among landholders within a community, and between them and conservation agencies attempting to influence their behaviour. In these relationships, there is an important role played by trust and the willingness of people to reciprocate positive behaviours. These insights go beyond the usual scope of thinking within agricultural extension, with its emphasis on building the human capital of individual community members, in terms of their awareness, knowledge, attitudes and skills. The research shows that success in fostering trust and reciprocity are important determinants of the success of community-based governance arrangements for natural resource management (NRM), which have been prominent in Australia over the past two decades. Community-based NRM governance can make a positive difference if implemented well, potentially enhancing landholder self-reliance and thereby accelerating the adoption and diffusion process for conservation practices that are sufficiently attractive to landholders. In particular, community-based NRM is most relevant to those conservation practices that are more effective when adopted widely rather than by just a few landholders. Examples include the establishment of native habitat in the form of vegetation corridors crossing multiple farms, or the management of groundwater to contain dryland salinity in regions where there is strong groundwater connectivity between farms. On the other hand, community-based NRM cannot work miracles. It is unrealistic to expect community-based arrangements to increase adoption greatly in cases where the conservation practices on offer are contrary to farmers' interests and thus not 'adoptable'.

The theme of not working miracles is continued in **Chapter 9**, in which Neil Barr discusses difficulties in the job of extension agents attempting to encourage adoption of conservation practices. Role conflict may emerge when extension agents are asked to promote the adoption of practices that are not in the best interests of landholders. It may be that short-term weather or market conditions have temporarily reduced the attractiveness of a practice, or it may be that the practice will remain unattractive in the long term. Extension agents are, of course, acutely aware of the fruitlessness of promoting such practices, but in some circumstances they are required by their funders and/or employers to attempt to work miracles. Such cases violate the implicit social contract between extension agents and farmers – the expectation that extension agents will provide farmers with advice that enhances their interests. This leads to a loss of credibility of the advisor, and a loss of trust in the relationship, reducing the effectiveness of extension efforts more generally. It means that the extension agent is consigned to a relatively minor, non-influential role in a farmer's decision-making process. Only highly credible, highly trusted individuals are admitted to the inner circle used by farmers to support their decision making, while less credible people are limited to merely providing information. This highlights the importance of understanding the likely adoptability of new practices prior to undertaking extension efforts to promote their adoption. For example, in some regions, the rapid growth in numbers of small farmers oriented towards life style rather than production has important implications for what can realistically be expected from promotion of particular practices. These insights point to the need for government agencies and other environmental organisations to develop a much more sophisticated approach to their thinking about adoption of new practices, one that is based on the understandings presented in the chapters in this book.

Complementing the discussion of the family farm business in Chapter 7, Cathy McGowan in **Chapter 10** focuses on the role of women in agriculture, arguing that extension agents seeking to promote change on farms should take a proactive approach to involving women in the design, participation and evaluation of extension and other rural programs. As well as doing the right thing by women, there are pragmatic reasons for taking such an approach. Given that many decisions are made collectively by some or all members of the farm family (as discussed in Chapters 2, 4 and 7) rather than by 'the farmer' as an individual, the involvement of women in extension is likely to accelerate the process of practice change. Almost all of Australia's farm businesses are run by farm families, and women constitute a major part of the agricultural workforce. Nevertheless, there is relatively little effort to design or implement extension programs in a way that recognises the needs and preferences of women. Many barriers to women's participation in agricultural programs have been identified, including male attitudes, women's self-confidence, family commitments, lack of female role models, and a lack of awareness of opportunities. Cognisant of these barriers, McGowan outlines a strategy for extension agencies to use to increase the involvement of women in extension activities.

The influences of demographic and social changes on rural practice change are mentioned in Chapter 9. The issue is considered in more depth by Allan Curtis and Emily Mendham in **Chapter 11** as they discuss the difficulties of moving from policy objectives to real changes in land management. The chapter outlines changes in the social structure of parts of western Victoria and the influence that these changes are having on landholder decisions, particularly those regarding the management of natural resources. These issues are discussed in the context of Australian government programs that provide funding for regional community-based bodies to encourage adoption of conservation practices by rural landholders (also discussed in Chapter 8). The research presented reinforces the challenges to government programs seeking to achieve high adoption of conservation practices. As would be expected, there was a positive relationship between the receipt of public funding by landholders and their adoption of conservation practices. Nevertheless, the importance of unfunded voluntary adoption was also emphasised – for the conservation practices studied, more than half the adopters received no financial support from government. In part, this may be attributed to the contribution that government programs make to increasing human capital and social capital (Chapter 8). Nevertheless adoption overall was lower than desired by program managers and indeed involvement by landholders had decreased over time due to due a severe drought (see Chapter 9). Clearly, government funding cannot necessarily overcome other factors that influence adoption decisions. One such factor is rural property turnover. In parts of the studied regions, the 'social landscapes' are changing rapidly, leading to a greater diversity of landholder types, and moving away from the past dominance of agricultural production. Curtis and Mendham consider that, in some areas, it is likely that up to 50% of rural properties will change ownership in the next decade, and point out that the new owners will be substantially different to the longer-term owners. These differences will affect land management, and will mean that government programs will need to incorporate a greater diversity of approaches that encourage adoption of conservation practices. From the point of view of these programs, many of the new 'lifestyle' landholders have stronger preferences for environmental outcomes than the landholders they replace. On the other hand, their objectives for their own properties (often emphasising aesthetic values) may not be consistent with program objectives. Further, they often lack the skills, knowledge, resources and time required to implement the conservation works being promoted (Pannell and Wilkinson 2009), and many are absent from the property for much of the time. Clearly, these changes have important implications for the design and implementation of government programs.

Issues related to policy for rural practice change are further discussed by David Pannell in **Chapter 12**. There are many different delivery mechanisms that policy programs may use to encourage changes in land management. Typically, programs rely on one or two mechanisms, and apply them across all projects. However, as discussed in other chapters, circumstances facing different landholders in different regions are highly heterogeneous, so the uniform application of a policy mechanism is highly unlikely to be the most effective approach. In particular, the adoptability of any particular conservation practice varies widely from farm to farm, due to differences in farmers' goals, soil types, capital resources, farming systems, family circumstances and other factors. As implied by many chapters in this book, a practice may be highly suitable for adoption on one property, but highly unattractive on another. This has major implications for the choice of policy mechanism. There are several policy responses that may be appropriate when the practices being targeted by policy are adoptable, and a different set of policy responses when they are not. For example, extension alone can be a sufficient tool to promote the adoption of a practice that is highly attractive to landholders, but will not be sufficient in cases where the practice is unattractive. Australian policy programs have generally not recognised this, and have persisted in using extension to promote the adoption of non-adoptable practices, resulting in disappointment and frustration all round. Heterogeneity also means that public investment in changing land management will pay off to different extents in different parts of the landscape. There has been an increasing emphasis in Australian policy on attempting to target conservation investment in ways that will increase the overall return on investment. This inevitably means that fewer landholders receive funding than was the case under programs that primarily emphasised the development of human capital (Chapter 8).

Conclusion

The issue of practice change by rural landholders is multifaceted, and invites study by researchers with varying perspectives. Despite this, a strong continuity and consistency of themes and messages emerges throughout the book. Some of the apparent differences in interpretation between, say, economists and sociologists, amount to little more than differences in language or semantics when analysed closely, while others are perhaps more a matter of emphasis than of content. Common themes are the need for appreciation of the heterogeneity of land managers, the need for policy and extension programs to be designed based on a stronger understanding of the adoption process, and the need for a more realistic assessment of the likely levels of adoption in different circumstances. In some cases, this could be as simple as recognising the fruitlessness of attempting to promote the adoption of practices that are not adoptable. A more sophisticated understanding of the many issues affecting landholder decision making, as presented in this book, would assist in the more effective design of policy, governance arrangements and extension programs and will lead to the greater achievement of social, environmental and economic outcomes.

References

Feder G and Umali D (1993) The adoption of agricultural innovations: a review. *Technological Forecasting and Social Change* **43**(3/4), 215–239.

Feder G, Just R and Zilberman D (1985) Adoption of agricultural innovations in developing countries: a survey. *Economic Development and Cultural Change* **33**(2), 255–297.

Knowler D and Bradshaw B (2006) Farmers' adoption of conservation agriculture: a review and synthesis of recent research. *Food Policy* **32**(1), 25–48.

Marra M, Pannell DJ and Abadi Ghadim A (2003) The economics of risk, uncertainty and learning in the adoption of new agricultural technologies: where are we on the learning curve? *Agricultural Systems* **75**(2/3), 215–234.

Pannell D, Marshall G, Barr N, Curtis A, Vanclay F and Wilkinson R (2006) Understanding and promoting adoption of conservation practices by rural landholders. *Australian Journal of Experimental Agriculture* **46**(11), 1407–1424.

Pannell DJ and Wilkinson R (2009) Policy mechanism choice for environmental management by non-commercial 'lifestyle' rural landholders. *Ecological Economics* **68**(10), 2679–2687.

SELN (State Extension Leaders Network) (2006) 'Enabling change in rural and regional Australia: the role of extension in achieving sustainable and productive futures'. A discussion document produced by the State Extension Leaders Network. http://www.seln.org.au.

Vanclay F (2004) Social principles for agricultural extension to assist in the promotion of natural resource management. *Australian Journal of Experimental Agriculture* **44**(3), 213–222.

Vanclay F and Lawrence G (1995) *The Environmental Imperative: Ecosocial Concerns for Australian Agriculture*. Central Queensland University Press: Rockhampton.

2

Understanding and promoting adoption of conservation practices by rural landholders

David J Pannell, Graham R Marshall, Neil Barr, Allan Curtis, Frank Vanclay and Roger Wilkinson

Summary

Research on the adoption of rural innovations is reviewed and interpreted through a cross-disciplinary lens to provide practical guidance for research, extension and policy relating to conservation practices. Adoption of innovations by landholders is presented as a dynamic learning process. Adoption depends on a range of personal, social, cultural and economic factors, as well as on characteristics of the innovation itself. Adoption occurs when the landholder perceives that the innovation in question will enhance the achievement of their personal goals. A range of goals is identifiable among landholders, including economic, social and environmental goals. Innovations are more likely to be adopted when they have a high 'relative advantage' (perceived superiority to the idea or practice that it supersedes), and when they are readily trialable (easy to test and learn about before adoption). Non-adoption or low adoption of a number of conservation practices is readily explicable in terms of the failure of these practices to provide a relative advantage (particularly in economic terms) or a range of difficulties that landholders may have in testing them.

Introduction

Scientists have been striving to discover and develop new conservation technologies and practices that reduce the extent or the consequences of land and water degradation resulting from extensive agriculture and other rural land uses. Environmental programs like the National Landcare Program, the Natural Heritage Trust and the National Action Plan for Salinity and Water Quality have sought to encourage landholders to adopt these conservation practices, mainly through information provision and social processes, but to some extent, through the payment of financial incentives.

Some conservation practices have been readily and widely adopted by farmers. Most of these primarily address on-farm issues, including lime application to treat acid soils, reduced tillage for reducing erosion and improving soil structure, and, in relevant regions, claying to overcome water-repellence. In other cases, adoption has been modest at best. For issues like dryland salinity and biodiversity loss, the response by landholders as a whole is clearly insufficient to halt degradation processes.

Some scientists and policy makers have expressed frustration at the observed levels of adoption, and expressed a desire to understand it. We propose that it is understandable based on the large body of literature that considers the adoption of innovations by farmers and other landholders. This paper provides a selective review and interpretation of what is known about the determinants of adoption of new practices by landholders, both conservation practices and other types. The vast amount of literature on adoption of innovations has previously been reviewed in general (e.g. Rogers 2003) and for agriculture (e.g. Feder and Umali 1993; Ruttan 1996), including in Australian publications (e.g. Guerin and Guerin 1994; Lindner 1987) and in the context of extension (Black 2000). There have been reflective papers outlining lessons from adoption research for the adoption of conservation practices (e.g. Barr and Cary 2000; Cary *et al.* 2002; Pannell 1999; Vanclay 1992, 1997, 2004) and various studies of the adoption of specific conservation practices in Australian agriculture (e.g. Cary and Wilkinson 1997; Curtis *et al.* 2000; Kington and Pannell 2003; Lockie *et al.* 1995; Ransom and Barr 1994; Sinden and King 1990; Vanclay and Lockie 1993).

A feature of the adoption literature is its disciplinary fragmentation. Relevant research is conducted under the banner of economics, sociology, psychology, health promotion, marketing, agricultural extension and anthropology. Despite differences in language and perspective, the general lessons of these different branches of work are broadly consistent and can be readily translated between disciplines.

This paper is distinguished from past reviews in its aim to present a cross-disciplinary consensus involving authors who come from several disciplinary backgrounds – economics (G R Marshall, D J Pannell), rural sociology (A Curtis, F Vanclay, R Wilkinson) and psychology (N Barr) – and who have knowledge of agricultural and environmental issues. This diversity of backgrounds allows us to present and integrate a broader range of perspectives than in previous reviews and overviews. The paper has been written with conscious avoidance of discipline-specific jargon, theories and ideologies in order to allow wide communication and wide applicability.

The intended audience for the paper is broad. The results and implications presented here are relevant to scientists and their funding sources, extension agents and funding bodies, policy makers, managers in government agencies, natural resource management bodies (such as Catchment Management Authorities), non-government conservation organisations and farmer organisations. Extension has been given a particular emphasis in natural resource management programs to date. We define extension broadly to include public and private sector activities relating to technology transfer, education, attitude change, human resource development, and dissemination and collection of information. We emphasise that publicly funded extension is just one information source among many that landholders use.

We have attempted to relate the review to the perspective of our intended audience, focusing on its concerns about adoption of conservation practices. We imagine that our target audience outlined above has an objective to enhance some conservation outcome(s), and that this would require changes in behaviour by landholders. The question is, what might influence or limit the achievement of such changes? This leads us directly to consider the ways landholders identify and deal with problems and opportunities, so we are fundamentally concerned with landholder perspectives as well, but the primary aim is to translate those perspectives for those on the other side of the fence.

We have tended to use the more encompassing term 'landholders' rather than 'farmers' as many rural landholders are not farmers. By 'farmers' we mean landholders who use their land to produce food and fibre as a significant share of their family income. Some of the evidence we present may relate specifically to farmers, and where this is so we use the more specific term. As discussed later, non-farm rural landholders differ from farmers in systematic ways, and some important differences in their adoption of conservation behaviours have been observed (e.g. Curtis and Robertson 2003).

The core common theme from several decades of research on technology adoption is that landholder adoption of a conservation practice depends on their expectation that it will allow them to better achieve their goals. If the landholder does not perceive that goals are likely to be met, adoption will certainly not follow. Goals vary widely between individual landholders depending on their circumstances and personal preferences, but may include economic, social and environmental outcomes. Adoption is based on subjective perceptions or expectations rather than on objective truth. These perceptions depend on three broad sets of issues: the process of learning and experience, the characteristics and circumstances of the landholder within their social environment, and the characteristics of the practice. These three elements are considered in detail in the following three sections. The last section discusses the implications of the review for various stakeholders: researchers, extension agents, and policy makers.

The process of learning and experience to inform adoption decisions

Adoption is a learning process with two distinct aspects (Abadi Ghadim and Pannell 1999). One is the collection, integration and evaluation of new information to allow better decisions about the innovation. Early in the process, the landholder's uncertainty about the innovation is high, and the quality of decision making may be low. As the process continues, if it proceeds at all, uncertainty is reduced and better decisions can be made (Marra *et al.* 2003). At least for relatively simple innovations, a landholder's probability of making a good decision – one that best advances their goals – increases over time with increasing knowledge of, and perhaps experience with, the practice. Viewed in this light, the adoption process is never completed, in the sense of eliminating all uncertainty. All options are continuously open to question and review as new information is obtained or circumstances change.

The other aspect of learning is improvement in the landholder's skills in applying the innovation to their own situation (Abadi Ghadim and Pannell 1999; Tsur *et al.* 1990). Most farming innovations require a certain level of knowledge and skill for them to be applied in practice, and there can be a wealth of choices in the method of implementation (e.g. timing, sequencing, intensity, scale). Through learning-by-doing, as well as by reading, listening and watching, the necessary skills can be established and enhanced.

This dynamic process has been broken down into stages or phases in a number of different (though similar) ways (e.g. Barr and Cary 2000; Lindner *et al.* 1982; Pannell 1999; Rogers 2003). One typical description of the sequence follows.

(1) *Awareness of the problem or opportunity.* In this context, 'awareness' means not just awareness that an innovation exists, but that it is potentially of practical relevance to the landholder. There has been relatively little research on the transition from ignorance to awareness. Gibbs *et al.* (1987) found that the time taken for different farmers in South Australia to become aware of the existence of new innovations varied markedly. For many farmers it amounted to years despite the presence of extension activities designed specifically to raise awareness.

(2) *Non-trial evaluation.* Reaching stage (1), the point of awareness, is a trigger that prompts the landholder to begin noting and collecting information about the innovation in order to inform the decision about whether or not to go to the next step of trialing the innovation. Conducting a trial incurs costs of time, energy, finance and land that could be used productively for other purposes. To be willing to trial an innovation, the landholder's perceptions of it must be sufficiently positive to believe that there is a reasonable chance of adopting it in the long run.

(3) *Trial evaluation.* Trials contribute substantially to both the decision making and skill development aspects of the learning process. If small-scale trials are not possible or not enlightening for some reason, the chances of widespread adoption are greatly diminished. Landholders will be cautious about leaping to full-scale adoption due to the risk that the innovation will prove a full-scale failure. Practices which are not readily trialable may still be adopted (rotary milking platforms are an example), but generally only after substantial information seeking, discussion, analysis and reflection.

(4) *Adoption.* Depending on the trial results, use of the innovation may be scaled up. Typically, adoption is not an all-or-nothing decision – there is a grey area between small-scale trialing and the eventual scale of adoption (Duncan 1969). Adoption is often a continuous process, and may occur in a gradual or stepwise manner, sometimes ending in only partial adoption (Wilkinson 1989). Landholders often change and modify the practice or technology to adapt it to their own circumstances. Indeed, such adaptation is often an important outcome of the trial process.

(5) *Review and modification.* As noted earlier, in one sense, trialing is never completed, as landholders continue to evaluate the performances of all their practices. However, as the scale of use of an innovation increases, the balance of reasons for using the practice shifts from mainly evaluation to mainly beneficial use. Even after adoption peaks, there is a continuous process of review and modification.

(6) *Non-adoption or disadoption* If external information or local trial results are not sufficiently encouraging (i.e. it appears that the landholder's goals will not be advanced by the innovation), the landholder will reject the innovation. If it is initially adopted but then, for example, economic circumstances change or a superior replacement technology or practice becomes available, use of the original innovation may be scaled down and eventually discontinued.

The knowledge that is developed through this process is held by the landholder and is likely to be, to some extent, unique to them. It will probably be based on a mixture of scientific information, personal experience, and cultural influences. Culture includes laws, social norms, ideologies and other human-devised factors that influence behaviour. The culture of landholders is the result of a rich history and it is dynamic, being continually modified by many factors.

The learning process is influenced by the characteristics of individual landholders, their families and broader social environments and by the characteristics of the innovation (see later sections of this chapter). Before conducting a trial, the landholder's assessment of a technology or practice relies strongly on information from outsiders. At this stage, social and information networks would be important influences on the decision to proceed to trial, but after trialing has commenced, personal experience gained in that way is likely to be the main influence on further decisions (Dong and Saha 1998; Marsh *et al.* 2000). This has implications for the role of extension to promote adoption, as discussed later.

There is no guarantee that a landholder's subjective beliefs will ultimately lead them to a final decision that is actually the one most likely to achieve their goals in the best possible way.

Lindner (1987) argued that final adoption decisions are usually correct in the sense that they do actually advance the landholder's goals, but we suggest that some conservation practices are less likely to conform to this generalisation than productivity-related innovations. This may be because some conservation practices are relatively complex or the benefits and costs of some conservation practices are not clearly observable (see the section later on characteristics of conservation practices).

One example of a prominent conservation-related learning failure is provided by Pannell *et al.* (2001). They noted that many landholders (as well as scientists and policy makers) came to believe that successful prevention of dryland salinity on a farm would generally depend on cooperation from neighbours. Although this is true in some cases, in many it is not. This learning failure would seem to be due to the difficulty of observing the salinisation process, most of which occurs underground with long lags between cause and effect. The point is that even after more than a decade of these farmers using the tools recommended for salinity prevention, their personal experience had not allowed them to converge on an accurate understanding of the impacts of the salinity-management tools. In the absence of readily visible connections between action and response, they were not able to observe the error in what they had previously been led to believe.

This example highlights that the decision-making process is imperfect. In general, decisions about land management are made without full information. Indeed, there is a trade-off between the costs of acquiring additional information and the benefits of improved decision making, and landholders must strike a balance. Even if full information were available, there are limits to human mental capacity, so people often use heuristics, or 'rules of thumb', to simplify their choices. For example, Ostrom *et al.* (1994) proposed that individuals faced with problems of collective action involving large groups use heuristics to learn about their complex decision situation. They argued that individuals, lacking both the information and cognitive capacity to calculate all future contingencies in order to select a strategy, adapt their heuristics sequentially as they learn about their situation, including about the other people sharing the problem.

In the rest of this paper, we discuss a variety of influences on adoption decisions, but it is noted that because of imperfect decision making and heterogeneity of circumstances, we are always discussing trends and tendencies, rather than deterministic relationships. Factors that enhance the learning process can accelerate the adoption process. These factors may relate to the flows of information between people (e.g. the strength of social networks; see next section) and to characteristics of the innovation itself (e.g. easy observability of trial results; see section after next).

Social, cultural and personal influences on adoption decisions

The previous section was couched in terms of solitary decision making by an individual. Decision making is often also a social process as the decision maker enlists the involvement of others in the decision-making process, or operates as part of a family team. Where a farm is managed by a family (less than 10% of farms were run by single operators in 2001; Australian Bureau of Agricultural and Resource Economics 2003), the process of decision making is made more complex by the interplay of family members. Although, for convenience, we will often refer to the (singular) landholder or farmer, the reader should bear in mind that for many decisions, particularly larger ones, the decision-making unit can be a team, so that individual perceptions and goals influence a consensus rather than leading directly to a decision.

The significance and complexity of the decision are important factors influencing how widely the information-seeking net is cast and the extent to which decision making is shared.

A decision to change to a new herbicide or change wheat variety is likely to be a relatively simple process. A decision with potentially significant personal impacts (such as changing farm enterprises) is likely to be a shared decision.

Phillips (1985) found that a typical dairy farmer may embark on anything up to 30 learning projects in one year. A landholder (or landholding family) has limited learning time, and each project must compete with the others for that limited time. A minor decision will receive minimal information time, sufficient to achieve an acceptable solution, which is not necessarily the best possible solution. When contemplating a major change to their farming system, the farmer will often have a hunger for information on the particular issue. The more serious the consequences, the stronger the need for information and for confidence about the outcomes. For more important decisions, the dairy farmers in Phillips' (1985) study sought information from up to 40 people. Weaknesses in the farmer's knowledge were remedied by seeking technical information from people who were seen by the farmer as experts. These could be other farmers, company representatives, stock agents, consultants or researchers. In this initial stage, judgment on the source of information and its credibility is often only cursory. Non-feasible alternatives are rejected, but any option or advice that may be useful is retained (Janis and Mann 1977).

Depending on their personal and family circumstances, the issues about which landholders are most concerned at a particular time may not relate to conservation, or any aspect of land management. A particular landholding family that would at other times welcome information about land conservation practices may have no time or energy for it in the midst of more pressing family issues. Extension activities will at best only reach those landholders who are in a position to be receptive at the time the activities are delivered.

Relative to the information-seeking stage, the next stage, evaluation of the worth of information, is often more socially shared. Information must be assessed against the objectives of the landholder and their family. The goals of landholder families or individuals are heterogeneous, and can include the following: (1) material wealth and financial security; (2) environmental protection and enhancement (beyond that related to personal financial gain); (3) social approval and acceptance; (4) personal integrity and high ethical standards; and (5) balance of work and lifestyle.

Many more specific objectives can be identified, although they generally relate to one or more of the five broad goals outlined above. Makeham and Malcolm (1993) listed the following goals common within the farming community: to survive and grow; to set and overcome challenges; to farm well and be recognised for this; to improve the physical state and appearance of the farm; to acquire extra land or to control a larger business for the future and for heirs; to have a reasonable but not profligate standard of living which compares reasonably with others in farming and society at large; to earn enough profit to be able to improve and develop the farm so as not to have to work so hard in old age; to achieve capital gain and increase wealth; to have good quality animals and crops in good condition; to reduce income tax; to have a satisfying rural way of life; to have children well educated; to have enough leisure, increasing over time; to be a respected member of the community; and to have enough money to pursue non-farm interests. Some of these goals are complementary, others are in conflict, so trade-offs are often necessary.

One issue of long-standing discussion and debate has been the relative importance of economic factors as drivers of adoption. The debate started early, with contributions by some of the first researchers in the area (Griliches 1957, 1960; Havens and Rogers 1961). To this day, economists tend to put greater emphasis on the influence of economic factors than do sociologists.

We note that the different views of economists and sociologists sometimes have more to do with language than with substance. For example, from within our own ranks have come papers that describe exactly the same factors as being 'social' (e.g. Vanclay 1986, 1992) and 'economic' (e.g. Pannell 1999). Economists tend to have a broad concept of what constitutes an economic benefit (e.g. including consideration of costs and benefits over the long term, risk, the cost of foregoing other opportunities, the value of keeping options open, resource degradation, farming-system issues, and non-financial benefits and costs). Economists may actually be considering factors that others consider to be non-economic, but interpreting them through an economic prism. In the same way, sociologists have a broad concept of what constitutes a social benefit (Vanclay 2002).

In our judgment, there are several important influences on adoption, and economic benefit (broadly defined) is one of them. Reflecting our combination of economic and social perspectives, when we say 'economic benefit', we mean the net economic benefit as perceived by the landholder, not as calculated by an economist. Often the (perceived) potential financial gain plays an important role (e.g. Cary and Wilkinson 1997), although sometimes it is counterbalanced by concerns over issues such as time, lifestyle or risk. Some farmers place the desire to make more money low on their list of priorities (Hawkins and Watson 1972; Presser and Cornish 1968; Vanclay 2004). For most, making money will not be their core goal, but it will be an important tool for achieving higher order goals such as a secure family lifestyle or keeping the farm property in the family (which means that economic return is still an important influence on their behaviour). Further, even landholders with a low emphasis on generating additional cash income are unlikely to be attracted to adoption of practices that would involve large economic losses (e.g. removal of woody weeds in some situations).

When an adoption decision has a potential to threaten the higher order goals, the process of decision making is much more likely to be socially shared. Dealing with risky decisions with important consequences is a stressful experience for most people. Most decision makers cope with the stress of uncertainty by seeking both further information and social or family support for decision making, particularly in the non-trial evaluation phase. The issues will not only be 'will this work?' but also 'will these people share responsibility for the decision?', and 'will they support me if it fails?'

The more difficult the decision, the more the decision maker will engage and re-engage with their personal support network and with other sources of information. The major decision will often be preceded by a series of smaller decisions to continue investing time, effort, and sometimes money, in continuing the decision evaluation. At each of these subsidiary decision points, the decision maker (or members of the decision-making team) may seek the advice and support of close contacts (Phillips 1985). Later in the process, social commitment and support will help maintain confidence in the uncertain stages of testing and early adoption. Peer expectations of continued commitment or personal support and encouragement will reinforce commitment and provide a buffer against setbacks (Janis and Mann 1977).

When adoption is viewed as a social process, it becomes clear that one should expect adoption behaviour to be influenced by the personality of the decision maker, their social networks, personal circumstances and family situation. It seems that in the empirical literature every measurable characteristic of farms and farmers has been found to be statistically related to some measure of adoption of some innovation (e.g. Rogers 2003). This reflects the heterogeneity of adoption study settings, the very large size of the literature, and the variable quality of empirical studies (as noted, for example, by Lindner 1987; Vanclay 1986). Vanclay (1986) particularly criticised the statistical analyses in many studies for failing to untangle the effects of multiple causal variables properly.

Personality may play a major part in the style of decision making used by landholders, although, because of measurement complexity, it has rarely been studied. One important personality trait is 'locus of control'. Individuals with a strong belief in their own ability to influence the circumstances of their lives are described as having an 'internal locus of control'. Persons with this personality trait are likely to experience less stress in decision making. The individual portrayed in John O'Brien's famous Australian poem 'Said Hanrahan' no doubt had an external locus of control ('If we don't get three inches, man, or four to break this drought, we'll all be rooned,' said Hanrahan, 'before the year is out') and may have been more troubled by stress during decision making. The limited research into farmer stress in Australia has shown that financial difficulty alone does not predict stress. Stress instead results from a combination of circumstances and the interpretation placed upon those circumstances by the individual. There is great variation in psychological propensity towards the experience of stress (Cary and Weston 1978; Weston and Cary 1979).

Economists study 'risk aversion', which is perhaps akin to a personality trait. Risk aversion describes an individual's tendency to take or avoid risks in their decision making. Empirical evidence indicates that farmers vary widely in their personal degree of risk aversion (Abadi Ghadim and Pannell 2003; Bardsley and Harris 1987; Bond and Wonder 1980). The more risk-averse a landholder is, the greater will be his or her tendency to adopt an innovation that is perceived to reduce risk (e.g. Shapiro et al. 1992) or to not adopt an innovation that is perceived to increase risk (e.g. Abadi Ghadim et al. 2005).

Another important personality trait is introversion/extroversion. Shrapnel and Davie (2001) and Shrapnel (2002) examined the personality profile of a sample of Queensland graziers. Of 14 general personality styles expected in the wider community, graziers were found to generally fall into a limited suite of five styles. 'Our findings indicate that they are indeed a special breed, with characteristic[s] that set them apart from members of an urban community' (Shrapnel and Davie 2001, p. 177). These characteristics include a tendency to introversion and discomfort within group situations. Although this work is formative, it provides an indication of why one-on-one relationships are likely to be preferred by many farmers over group settings in the evaluation of options for important decisions. This personality trait will influence the extent and nature of a farmer's personal networks. Personal networks are an important influence on adoption behaviour and are increasingly important as a medium for the implementation of government and industry programs.

A widely discussed and long-standing concept is categorisation of people across a spectrum from innovators to laggards, presented with little change from Rogers (1962, pp. 168–171) to Rogers (2003, pp. 282–285). People do indeed have personal characteristics that influence their adoption decisions fairly consistently. However, the concept of adopter categories suggests that innovativeness is a personal characteristic that people apply equally to every adoption decision that they make. This is not so. People who adopt one innovation early are not necessarily early adopters of all innovations. It may be that the innovation in question is particularly attractive in their individual circumstances, whereas the same decision maker when considering a different innovation that is less attractive to them than to others may behave as a slow adopter or non-adopter.

Several aspects of the links between landholders and others may affect the adoption decision:

(1)   The existence and strength of landholders' social networks and local organisations (e.g. Sobels et al. 2001) and membership of organisations such as catchment groups have been shown to be positively related to adoption (e.g. Kington and Pannell 2003). A number of studies have found a positive relationship between membership of Landcare groups and adoption of some conservation practices (Cary et al. 2002; Curtis 1997; Curtis and De Lacy 1996; Mues et al. 1998), although the direction of causality is not clearly established.

(2)   The physical proximity of other adopters is positively related to adoption (e.g. D'Emden *et al.* 2006; Hagerstrand 1967; Ruttan 1996).

(3)   The physical distance of the property from sources of information about the innovation is important – more distant landholders are less likely to adopt, perhaps because the information appears less relevant to them than to those who are close to the information source, or perhaps because they receive less exposure to the information (e.g. Lindner *et al.* 1982).

(4)   A history of respectful relationships between landholders and advocates for the innovation, including scientists, extension agents, other landholders, and private companies, is positively related to adoption, through enhanced trust in the advice of the advocates (e.g. Anderson 1981; Marshall 2004a, 2005).

(5)   Ethnic and cultural divisions within a landholder population can act as significant barriers to the flow of information about environmental innovations (Stoyles 1992).

(6)   Extension, promotion and marketing programs by government workers and/or the private sector can be positively related to adoption (e.g. Llewellyn 2002; Marsh *et al.* 2000). Characteristics of extension agents that enhance effectiveness of extension are discussed in the Implications section later.

Demographic and situational variables are judged to be important because they will influence the goals of the landholder and potentially influence the capacity to adopt an innovation. Some examples of these variables are listed below.

(1)   Cary *et al.* (2001) found that profit expectations are an important influence on investment plans (and, thus, on adoption decisions). Lack of financial viability would be expected to inhibit adoption of innovations by reducing the capacity to adopt, rather than the benefits of adopting. Cancian (1979) conducted a meta-analysis of the relationship between income and adoption and concluded that it may not be linear.

(2)   Access to and reliance on off-property income can influence the adoption of practices by increasing financial security but also by decreasing the tendency to adopt some practices that would increase profitability but involve greater management demands (Kebede 1992).

(3)   Property size is often, but not always, related to innovation adoption (e.g. Abadi Ghadim *et al.* 2005) – larger areas tend to increase the overall benefits of adoption of beneficial innovations and so increase the likelihood of adoption. Alternatively, social issues related to adoption may also lead to adopters having larger properties. In north-east Victoria, conservation cropping technology was more likely to be owned by operators of larger and specialist cropping enterprises rather than owners of smaller or opportunistic cropping enterprises (Wilkinson and Cary 1993). In north-central Victoria, the adoption of perennial pastures was strongly related to property size, but the adoption of tree planting was not (Wilkinson and Cary 1992). D'Emden *et al.* (2006) also found a lack of relationship between farm size and adoption of conservation tillage in Western Australia.

(4)   Age would appear to be of particular relevance to adoption of conservation practices that have long lags between investment and payoff. If a farm is not to be passed on to the farmer's children, and if the benefits of conservation practices are not expected to be fully reflected in the sale price of the farm, then older farmers may have less incentive to invest in something that will be primarily of benefit to the subsequent owner (Gasson and Errington 1993). We speculate that age may also influence adoption via a correlation with physical health. However, the evidence of a relationship between adoption and age, stage of life or experience is mixed. The most extensive meta-review of socio-economic factors influencing adoption found both positive and negative relationships between age and

adoption (Rogers 2003). The limited research addressing the influence of age on adoption of conservation practices (e.g. Cary *et al.* 2002; Curtis and Byron 2002; Latta 2002) is just as mixed.

(5) There can sometimes be relationships between education and the adoption of conservation practices. It has often been concluded that beneficial innovations tend to be adopted more quickly by landholders with higher levels of education (e.g. Feder *et al.* 1985; Goodwin and Schroeder 1994; Kilpatrick 2000; Rahm and Huffman 1984). However, in the case of a complex technology or practice that is actually disadvantageous when all of its effects are considered, education may tend to reduce or delay adoption by allowing the limitations of the practice to be recognised (e.g. Marsh *et al.* 2006). These limitations may go unrecognised by less educated landholders, who consequently adopt the practice mistakenly. Kilpatrick (2000) has shown the catalysing impact of education in general (not just agricultural education and training) on farmers' abilities and levels of interest in modifying farming practices. Given the decline in the traditional family farm apprenticeship as a means of entering farming in preference for a longer period of education (Barr 2004), the future farm management force will be increasingly educated and, presumably, increasingly interested in ongoing self-education about farming systems. Nevertheless, we suggest that a farmer's general level of education is likely to be less important as a predictor of adoption than their participation in specific relevant training courses.

(6) The reason for holding land (e.g. agricultural production *vis a vis* lifestyle) can influence adoption decisions. As Vanclay (2004, p. 214) observed, 'Different farmers have different priorities, different understandings, different values, different ways of working, and different problems'. Regions within comfortable driving distance of major cities and regional centres in some Australian states (particularly New South Wales and Victoria) have seen social and demographic changes resulting from city dwellers purchasing what was formerly extensive farming land and pursuing their rural dreams. In these regions, traditional commercial agriculture has become a less important land use than it once was, occupying a declining proportion of the land, and the trend in this direction will continue (Barr and Wilkinson 2005). The new landholders may not have the time or financial resources for investment in large-scale adoption of expensive conservation practices (e.g. establishment of woody perennials) even if such practices would be financially beneficial in the long run (Nicoll 1994). Vanclay *et al.* (1998) promoted the idea of 'farming styles' as a useful mental model for making some sense of this diversity of farming objectives. The farming styles approach has been used in an exploratory fashion to try to explain conservation innovation adoption with mixed success (Howden *et al.* 1998; Howden and Vanclay 2000; Mesiti and Vanclay 2006; Vanclay *et al.* 2006).

Attributes of practices that affect adoption innovations

We consider that there are two broad categories of characteristics of a technology or practice that drive its adoption or non-adoption: its relative advantage and its trialabilty. 'Relative advantage' refers to the perceived net benefits if you do adopt, while trialability refers to how easy it is to move from non-adoption to adoption, via a learning phase. Other characteristics that are mentioned in the literature will be discussed under these broad headings.

Relative advantage

Relative advantage means 'the degree to which an innovation is perceived as being better than the idea [or practice] it supersedes' (Rogers 2003, p. 229). Relative advantage depends on the landholder's unique set of goals and the biophysical, economic and social context where the innovation will be used. Relative advantage is the decisive factor determining the ultimate level of adoption of most innovations in the long run.

Relative advantage depends on a range of economic, social and environmental factors, such as:

(1) *The short-term input costs, yields and output prices of the innovation or of other activities that it affects.* For example, Marsh *et al.* (2000) and Abadi Ghadim *et al.* (2005) found that the short-term profitability of new legume crops (e.g. lupins and chick peas) significantly influenced their adoption. Sinden and King (1990) and Cary and Wilkinson (1997) found that short-term expectations about variables related to profitability influenced the adoption of conservation practices.

(2) *The innovation's impact on profits in the medium to long term.* The relative importance of short-term and long-term profits depends on the individual's personal goals and circumstances, but most farmers profess concern with outcomes beyond the short-term (Makeham and Malcolm 1993; Wilkinson and Cary 1992).

(3) *The innovation's impacts on other parts of the system within which it will be embedded.* For example, a legume crop or pasture can increase the yield of subsequent cereal crops by nitrogen fixation and impacts on crop disease (e.g. Pannell 1996). Long-lived trees may reduce the flexibility of crop producers who wish to switch in and out of crop production from year to year in response to weather conditions – an important strategy in some low-rainfall environments (Cary 1986; Kingwell *et al.* 1993).

(4) *Adjustment costs involved in adoption of the innovation.* One reason that the adoption of integrated pest management has been relatively slow is the relatively high adjustment cost it entails (Wiebers 1992). Of course, adjustment costs can be borne if the attraction of a practice is strong enough, as has widely occurred with conservation tillage.

(5) *The innovation's impacts on the riskiness of production* (Abadi Ghadim *et al.* 2005; Marra *et al.* 2003). The relative advantage of an innovative land use would be reduced if it were perceived to be more subject to price variability, to establishment failure, or to yield losses due to drought, weeds or pests than the current land use.

(6) *The innovation's compatibility with a landholder's existing set of technologies, practices and resources* (Kaine and Lees 1994). For example, a new higher yielding wheat variety is readily adoptable by an existing wheat farmer because it is compatible with the farmer's current machinery, rotations, agronomic practices, herbicide usage, and so on. However, to such a farmer, a new type of tree crop is unlikely to be as compatible with existing practices, so the cost of making the transition to a new farming system that includes the tree crop would tend to reduce its relative advantage and moderate its adoption. Some practices are sensitive to the soil types on which they are used, and so may have higher relative advantage on some farms with particular soils. For example, different crops prefer sandy soils or loamy soils. Some plants are sensitive to soil acidity or soil salinity and some are tolerant, influencing their attractiveness to a farmer depending on which soils are present on the farm. One resource that is a critical determinant of a farmer's ability to make an innovation work is their own management skill.

(7) *The innovation's complexity* (Rogers 2003; Wilkinson 1989). Complexity may increase the intensity of effort required for ongoing management, and the risk of the innovation failing in any given year, each of which reduces the innovation's relative advantage. Alternatively, an innovation may be no more complex in itself, but adoption of it may add to the overall complexity faced by the land manager. For example, a farmer considering a suite of crop types may find that the managerial complexity of managing five different

crops on the farm, each with its own requirements for machinery, agronomy, marketing, and storage, is unacceptably greater than the complexity of managing four. Even if the fifth crop type would actually be profitable to adopt, this could be outweighed by considerations of inconvenience, stress and risk. Likewise, the conversion from an annual pasture production system to one including perennial pastures such as lucerne can entail significant flow-on impacts upon the farm system.

(8) *Government policies.* Relative advantage can be affected positively or negatively by government policies. For example, in the US, support programs that are based on yield tended to increase the relative advantage of the intensification of farming and thus increase adoption and use of herbicides (Helms *et al.* 1987; Miranowski *et al.* 1991).

(9) *The cost or profitability of the traditional practice which the innovation would replace.* For example, increases in the price of fuel and labour tended to increase the adoption of herbicides in the US, as herbicides substituted for cultivation which was becoming more costly (Carlson and Wetzstein 1993; Miranowski and Carlson 1993). As another example, D'Emden *et al.* (2006) found that the adoption of conservation tillage in Australia has been significantly affected by the ratio of glyphosate to diesel prices. They estimated that the time to adoption was halved because of the fall in the relative price of glyphosate over the study period. Further, a history of specialisation in a particular practice is likely to increase the farmer's skill level and managerial abilities specific to that practice, and so reduce the financial relative advantage of an alternative innovation.

(10) *The compatibility of a practice with existing beliefs and values.* At least in the short to medium term, farmers may consider themselves to be wedded to production of a particular output (e.g. grain cropping, not livestock or trees) or to a particular method of production (e.g. traditional farming methods, not organic) because they identify with it personally (e.g. all my friends are wheat farmers, I am a wheat farmer too; it is what I like doing, it is what I'm good at, it is what my family does, it is an important and respectable occupation for me). This reduces the relative advantage of alternatives.

(11) *The impact of the innovation upon the family lifestyle.* The failure of marital relationships is one of the major causes of farm business failure (Barr 1999). To most farming families, the farm is the means towards the goal of a secure family life. Some innovations can cause a change in the quality of the family lifestyle or potentially even marriage relationships, and so their relative advantage can be strongly affected by this. In the eastern wheatbelt of Western Australia, the large-scale introduction of perennial pastures into a cropping system may threaten the traditional summer holiday that allows the family to temporarily escape the harsh summer wheatbelt environment. The adoption of redesign of the farm irrigation system in northern Victoria was in part assisted by the advantages of less broken sleep and more time for couples to share in the evening.

(12) *Self-image and brand loyalty.* Relative advantage can be affected if an innovation changes the social standing of people within the local culture. In some situations this can accelerate the rate of adoption while in others it can retard adoption. The adoption of cross-breeding of beef cattle was slowed in the high-rainfall beef zone by the traditional status gained by the production of pure-bred Hereford cattle. Although minimum-tillage techniques are now widely adopted, early adopters tell stories of the social challenges of not cultivating while neighbours were busy cultivating and maintaining a social network through tractor-cabin CB radios. The social stigma of having an 'untidy' farm (one without straight furrows) also delayed the adoption of zero-till (Coughenour and Chamala 2000).

(13) *The perceived environmental credibility of the practice.* We would expect environmental credibility to enhance the relative advantage of a practice. Environmental advantage is

not always clearly observable, as recent changes in the understanding of dryland salinity attest (Ridley and Pannell 2005). In north-central Victoria, trees and perennial pasture have differed in their environmental credibility for dryland salinity management. Tree planting by farmers was based on 'symbolic' beliefs (symbolising a personal expression of concern for the public good), while pasture sowing was based on 'instrumental' beliefs (providing tangible personal benefits such as increased production) (Cary 1993). Consequently, farmers planted a similar number of trees no matter what size their farm, while those with large farms tended to sow a larger area of perennial pasture than those with small farms (Wilkinson and Cary 1992).

The crucial role of 'relative advantage' as a driver of adoption, and the importance of profit as one of the drivers for most farmers, has strong implications for conservation practices. Among those farmers with a focus on profit, the farm-level economics of a proposed conservation practice will be important. Those conservation practices that are not profitable at the farm level will tend to be adopted only by farmers with stronger conservation goals. The lower the perceived profitability, the stronger the conservation goals need to be for adoption to occur. Unprofitable conservation practices are likely to be more widely adopted if they are able to generate conservation benefits when adopted at a small scale. Conservation land uses that require adoption at large scale to generate conservation benefits will probably not be adopted sufficiently if they are perceived to be less profitable than the land uses they replace.

There are numerous examples that illustrate and reinforce these implications. There has been a very widespread but small-scale adoption of unprofitable conservation practices among many landholders, triggered in part by government programs such as the National Landcare Program and the Natural Heritage Trust (Mues *et al.* 1998). For example, the resources committed to fencing off streams and remnant vegetation attest to the conservation concerns of many landholders. However, for all but a minority of farmers, the costs and areas involved are small relative to the scale of the farm businesses (e.g. Curtis and De Lacy 1996).

Some conservation-related practices have been adopted very widely and over very large areas in Australia, most notably reduced tillage and liming of acid soils (e.g. Mues *et al.* 1998). These are both practices that contribute positively to farmers' economic goals in the medium term in many locations. This highlights that the relative advantage that drives adoption may not necessarily relate to the environment. Indeed, environmental benefits can often be most readily achieved by developing conservation practices that provide a commercial advantage to farmers.

Conversely, the scale of adoption of perennials for salinity abatement in low- to medium-rainfall areas has been much less than needed to significantly reduce the salinity threat (e.g. ABS 2002; Kington and Pannell 2003). A recent comprehensive review of the economics of salinity abatement measures available to grain growers provides a convincing explanation for this, as there were few examples of locations and practices where the economics favoured high levels of adoption (Kingwell *et al.* 2003).

Other factors that tend to reduce the relative advantage of at least some conservation practices are as follows.

(1) *High establishment costs.* Land conservation practices are often characterised by high up-front costs, and benefits that occur some time in the future (e.g. the establishment of woody perennials for salinity mitigation). Large up-front costs plus accumulated interest therefore reduce the attractiveness of these practices.

(2) *Long time scales.* A number of the land degradation issues of concern are long-term by nature. In some cases, degradation processes of concern occur over decades (e.g. dryland salinity, soil acidification, decline of remnant native vegetation) and the practices

designed to ameliorate the degradation can be slow to take effect. For this reason, land conservation innovations are often particularly susceptible to the problems of up-front costs and delayed benefits, as outlined above. It also means that those landholders who are forced by circumstances to give priority to short-term profits are unable to adopt even if the innovation would eventually generate benefits sufficient to offset the up-front costs plus interest.

(3) *Riskiness.* Long time lags between planting and harvest (e.g. decades for many woody perennials) contribute to the riskiness of production, because they give scope for unanticipated developments in product markets, development of competing technologies, accidents or natural events (e.g. fire or pests) to damage the harvestable product.

(4) *Complexity.* Some conservation practices are relatively complex, further reducing their relative advantage. For example, many farmers perceive that lucerne pasture (a perennial pasture plant that helps in salinity management) requires a greater intensity of management than do traditional annual pastures (e.g. Lodge 1991). The prospect of expanding the portfolio of farming activities by adopting additional land uses for conservation purposes is likely to be perceived as increasing the overall complexity of the land management system. Greater complexity may contribute to an increased risk of failure.

(5) *Spillovers.* Some of the benefits from conservation practices can extend to individuals other than the adopting landholders. A result can be that adoption is less than it would be if the landholder considered all the benefits to the broader community. However, there is scope in some cases for landholders to reciprocate part or all of the benefits they obtain from one another's adoption, and, thus, narrow the gap between the interests of potential adopters and the wider community. An example is shallow groundwater pumping to lower the watertable beneath adjoining irrigation properties. In such a situation, adoption of pumping on any of the properties would lower the watertable not only on that property but, to a lesser extent, on the neighbouring properties as well. If a landholder expects the neighbouring landholders to reciprocate his or her investment in pumps, the incentive for the first landholder to invest is increased. However, the expected relative advantage to any landholder from adopting will be lower if the reciprocation cannot be trusted (i.e. if the other landholders are expected to 'free ride' on the actions of the first landholder) (Marshall 2004b, 2005).

Given that it can be difficult for landholders to develop mutual trust, a perception that adoption of a conservation practice generates reciprocal spillover benefits can itself be a disincentive to adoption, even if the perception is incorrect. As an example where this occurred, in the 1980s and '90s, there developed a widespread misperception among Australian farmers that adoption of measures to ameliorate dryland salinity on farms generates reciprocal spillover benefits, although this is rarely true in reality. For many farms (particularly in Western Australia), the control of salinity is solely dependent on actions within the farm boundary (Pannell *et al.* 2001). In all states where spillovers from dryland salinity are significant, they are almost always unidirectional (uplands affecting lowlands) rather than reciprocal.

Trialability

In the previous section we discussed a number of social, cultural and personal factors that influence learning about an innovation. Here we consider characteristics of the innovation itself that affect how easily the landholder can learn about its performance and optimal management – in other words, the trialability of the innovation. This does not merely refer to the ease of physically establishing a trial, but encompasses factors that influence the ability to learn from a trial, such as the complexity of the issue being addressed.

Trialability has been found to enhance adoption (e.g. Ohlmer *et al.* 1998). As noted earlier, trialing an innovation provides information that reduces uncertainty about the relative advantage of the practice (Tonks 1983). Thus, trialing is important because it can increase the probability of the landholder making a correct decision. Trialing also provides an opportunity for the landholder to learn the skills needed to apply the innovation. The small-scale nature of a trial allows the landholder to avoid the risk of large financial costs if the practice turns out to be uneconomic or fails due to inexperience. The reductions in uncertainty and risk from these two aspects are themselves of benefit to the majority of people who are psychologically averse to risk and uncertainty.

The trialability of a given practice is affected by a number of factors, including those listed below. Note that several of these factors were also listed as influences on relative advantage. These factors influence adoption through both channels.

(1)   The divisibility of an innovation refers to its use on a small scale, or the use of a sub-component of an innovation package. A degree of divisibility is essential to allow small-scale trialing for learning purposes (Leathers and Smale 1991). For example, a new herbicide would be trialable on a very small scale. In contrast, a new land use intended to contain a rising watertable requires a minimum scale for its effects to be apparent and hydrological evidence indicates that the necessary scale for impacts to be apparent at any distance from the trial is very large (e.g. George *et al.* 1999). Indeed, it appears that almost full adoption is often necessary. High fixed costs for an innovation (e.g. the need to purchase new machinery) reduce its divisibility. Even when an innovation package is promoted to farmers as a tightly-bundled, complex technology, farmers have a strong propensity to pull it apart and adopt only some of its components, or adopt selected components in a stepwise manner (Wilkinson 1989).

(2)   The observability of results from an innovation is positively related to adoption (Pannell 2001b), at least in part, due to its influence on trialability. Trialing a practice becomes less costly, and thus more likely to be seen as worthwhile, the greater the observability of trial outcomes. Higher observability means that fewer trials may be necessary to sufficiently reduce uncertainty to make the choice between adoption and non-adoption. Observability also enhances the prospects of 'over the fence' learning by landholders, and, thus, promotes diffusion of a practice (Geroski 2000; Shampine 1998). The impacts of a new production-oriented practice (e.g. a new herbicide) are often readily and rapidly observable. In contrast, a new land use intended to control a rising watertable in a neighbouring paddock may have effects that are long delayed and physically difficult to observe, requiring the costly installation of piezometers. Even when a piezometer reading is obtained, given the considerable complexity and heterogeneity of underground geological structures in agricultural regions of Australia, it can be difficult to know how representative the observation is. Perhaps in recognition of the importance of observability, there have been a number of initiatives attempting to increase the observability of watertable rise (e.g. Watertable Watch bore flags in Victorian irrigation settlements in the 1980s). There is also considerable difficulty in attributing any change in a watertable to the practice that is being trialed. One difficulty is the absence of a suitable control against which the result can be compared. When trialing an innovation, such as a new grain crop variety, it is relatively easy to compare the crop's performance with traditional varieties in the same growing conditions. When trialing an agronomic practice, results can easily be compared with and without the practice. However, for a perennial plant enterprise established in order to prevent rises in the watertable, such comparisons are much more difficult and less informative.

It may be that the only available method for assessing the impact of an area of perennials on the watertable would be to observe the deviation from a previously recorded trend. The need to look for a deviation in a historic trend, rather than comparing two current treatments, would add to the delay before any conclusion could be reached with confidence. This is because, in the absence of a control treatment, it is more difficult to determine whether any observed deviation is attributable to the new practice or to other factors, such as atypical rainfall. Uncertainty about the response lag time adds to the difficulty of interpreting any observed trend deviation.

(3)   The longer the lag, the less trialable is the innovation. For example, even if observability was high, and a control treatment was available for comparison, groundwater movements are slow, so it may be a long time before a landholder's uncertainty about the soundness of a practice for groundwater management is sufficiently reduced to prompt widespread adoption. Slowness reduces the overall value of trialing and may contribute to a judgment that the benefits of the trial do not outweigh the costs.

(4)   The complexity of an innovation is negatively related to its trialability, and eventual adoption. The greater the complexity, the greater the information that landholders require to be certain about the consequences of adopting it. In a sense, this aspect of complexity is related to the innovation's observability. If there could be multiple consequences from adoption, it is more difficult or expensive to observe them all, and landholders may take a longer duration of trialing to develop confidence in their judgments. A separate issue is that greater complexity of an innovation will likely increase the risk of technical failure when conducting the trial. In general, greater complexity increases the difficulty, required effort and time to learn about the innovation's performance and how best to implement it. This reduces the anticipated value of trialing and so may discourage it from occurring, to some extent.

(5)   The cost of undertaking a trial will be negatively related to adoption. For example, if seed of a new crop is in short supply, and so temporarily expensive, farmers may decide to delay their trialing of the crop. Where a large-scale trial is necessary (e.g. for water management technologies), the cost of trialing is correspondingly larger, and it is therefore less attractive. A large trial consumes land, labour and finance which could otherwise be used productively. In these ways, high costs reduce trialability.

(6)   Threats to a biological trial may include drought, diseases, pests, and establishment failure. Trials of any innovation always face a risk of failure, but given the large area for which a trial of a tree enterprise appears necessary to discern watertable impacts, the potential losses from a trial failure are relatively large. This provides further discouragement to a risk-averse landholder considering such a trial.

(7)   For the information from a trial to have value for decision making, the trial needs to be indicative of the innovation's performance in the long run. If the technology or practice used in the trial is implemented poorly, then the trial will clearly be less likely to meet this requirement. Poor implementation is more likely when the innovation is radically different from practices with which the landholder is familiar, and this does appear to describe the situation for some conservation practices.

(8)   Similarity in behaviour of the innovation to a familiar practice can be helpful in the learning process, and so, in a sense, can enhance trialability. For example, if the pattern of yield responses to weather for a new crop is similar to that for familiar crops, a farmer can extrapolate more readily from a small number of observations of the new crop. Abadi Ghadim *et al.* (2005) found that differences among farmers in their perceptions about the similarity of responses between yields of a new crop (chick peas) and a traditional crop

(wheat) was a significant variable explaining their adoption intentions. New conservation-related land uses are likely to be less similar to traditional land uses in their behaviour than are new production-oriented technologies.

(9) The presence of spillover effects can reduce the motivation for trialing. In a survey of farmers in the upper Kent River catchment of Western Australia, Kington and Pannell (2003) found that 62% of farmers believed that their neighbours were causing spillover salinity impacts. Although the survey did not explore the proportion of the problem that was attributed to inter-farm flows, it appeared that many of these farmers were significantly overrating the extent of the spillover problem (Pannell *et al.* 2001). If farmers believe (rightly or wrongly) that the rise in their watertable is due to the management practices of their neighbours, their motivation to trial a practice for groundwater control on their own farm is reduced.

Implications

Implications for research

Following Marsh (1998), we provide the following suggestions for biophysical scientists to help them achieve greater adoption by landholders of conservation practices being researched.

(1) Be conscious of the type of practices that landholders adopt more readily – those with high relative advantage and high trialability. Appreciate that landholders have legitimate reasons for non-adoption (Vanclay 2004). If the community has a wish to reduce a particular form of environmental degradation originating from rural properties, but the available practices for reducing the degradation conflict with goals of landholders (e.g. salinity treatments highly unprofitable to farmers), one sound response for scientists is to consider the viability of developing new technologies or practices that achieve both community and landholder goals.

(2) Encourage a participatory process. Working with landholders forces researchers (and extension workers) to recognise that their own goals may be different to landholders' goals, and reduces the risk of them making incorrect or over-simplified assumptions about what landholders' goals really are. In a participatory project, the research/extension can be adapted in response to this improved understanding. Such interaction also increases landholders' knowledge of the research and their ownership of, and faith in, the results. It may help landholders to understand and appreciate the goals of researchers. Participation also helps to develop better programs and recommendations by making better use of local knowledge so that recommendations are more often corroborated by subsequent experience, and in this way promotes landholders' trust in research, development and extension over the longer term.

(3) Look constructively at what landholders are doing already. Work with them where possible rather than against them (or at least acknowledge the difficulty of getting them to stop believing that what they are already doing is appropriate). This suggestion acknowledges the importance of local knowledge in landholders' decision making, and the importance of respecting their personal goals and perceptions. We suggest that scientific and local knowledge can be highly complementary.

(4) As we have argued, adoption of conservation practices by landholders is not solely a biophysical issue, it is also an economic, social and psychological issue, so biophysical researchers can benefit from working closely with economists, sociologists and psycholo-

gists. Social scientists should be involved in projects from an early stage, including in problem definition and project design, so that their advice can influence the direction of the research, rather than being limited to analysing the results (e.g. attempting to explain landholders' responses or lack of response).

Attending to these suggestions would help to enhance trust and credibility in the relationship between researchers and landholders. This is crucial if researchers are to influence the adoption process.

Kaine and Lees (1994) suggested the idea of research and development market segmentation on the basis of farming contexts, and Kaine *et al.* (2005) proposed the same idea for extension. Mesiti and Vanclay (2006) used a farming styles approach as an alternative to market segmentation. Market segmentation may be less straightforward in agricultural research and extension than in retailing, where variables like age, education and income have proven useful. The most useful variables in differentiating market segments among landholders are often psychological rather than demographic, and hence are more difficult to observe. Nevertheless, in practice, agricultural and environmental research and development does focus on locally relevant issues to a considerable extent. Kaine and his co-authors use an approach where the focus is on the farm context itself and the fit of the innovation within that context.

Given the importance of trialability for adoption of an innovation, it may be useful for researchers and extension agents to consider ways in which landholder learning from trials can be enhanced. One possibility suggested by Abadi Ghadim *et al.* (2005) is to provide information about the trial performance of familiar reference land uses or practices that are as similar to the innovation as possible, in conjunction with information about the performance of the innovation. It may be feasible to facilitate physical observation, or at least present results of physical measurements, of important processes that are not readily visible (e.g. groundwater processes). Perhaps it is possible to provide rules of thumb about final yields based on the early growth rates of plants that have long lags before harvest (e.g. woody perennials). Similarly, where a novel land use requires large-scale adoption to achieve environmental benefits, ways to predict those benefits based on performance in small-scale trials may be helpful.

Implications for extension

A criticism of traditional extension is that it viewed the extension process primarily as a matter of communication. Lack of adoption was blamed on a failure of the extension communication process. The solution was to target extension more effectively and to improve the methods of information delivery. The assumption was that farmers were information deprived and relatively passive recipients of knowledge. In reality, farmers have excessive information (e.g. from consultants, banks, accountants, agronomists, agribusiness firms, other landholders), some of which is conflicting, and they are almost never passive recipients. Recognising its place within this complex web of information sources, extension needs to be more focused on credibility, reliability, legitimacy, and the decision-making process. Features of current conservation-related extension that mitigate against the development of credibility include: short-term funding, rapid turnover of staff, the youthfulness and inexperience of many staff, and the lack of technical farming expertise of many staff (Vanclay and Lawrence 1995).

Expectations for extension

Even with the most expert and persuasive extension, landholders are not likely to change their management unless they can be convinced that the proposed changes are consistent with their goals. Therefore, expectations about the extent of change that is likely to result from extension need to be realistic (Vanclay 2004). Large changes made by large numbers of landholders are not likely to be attributable to extension in most cases, partly because landholders and their lands are highly heterogeneous. Any given practice advances the goals of only some landholders, and often only on some of their land.

It is likely that the main contributions of extension will be through raising awareness and, to some extent, changing perceptions of the relevance and performance of an innovation. It is much more difficult (and sometimes ethically contentious) to change people's goals. It seems that the Landcare movement in Australia has increased the emphasis given to conservation goals by landholders, but the extent of increase has been modest for most landholders (Reeve 2001) and the movement has, perhaps, reached the limits of its influence (Reeve *et al.* 2002).

This suggests that for many innovations, extension's main role will be to accelerate the adoption process, rather than to lift the final level of adoption (Marsh *et al.* 2000). Exceptions to this may include practices which would have entirely failed to diffuse in the absence of extension, perhaps due to problems with trialability (e.g. low observability, high complexity). Related to this, extension is unlikely to persuade landholders to make greater use of a practice with which they already have personal experience, unless the extension provides new information about a change that increases the attractiveness of the innovation (e.g. new information about how to better implement the innovation, or about new incentive payments to encourage adoption).

Another important issue for extension (as for science) is that it does not have automatic legitimacy and credibility (Vanclay 2004) – these have to be earned. A landmark study of the social process of agricultural extension in the 1970s showed that the key determinant of an adviser's credibility to a farmer was trust. Trust was, in turn, strongly related to the extent a farmer believed an adviser understood and respected the goals of the farmer (Anderson 1979, 1981). Trust determines the nature of the role that an adviser may play in the social aspects of the decision-making process of the landholder. Without trust, an adviser may only expect to participate as a provider of information that will be later evaluated within a closer circle of trusted contacts (Phillips 1985). The adviser who is trusted may be invited to participate at a deeper level of decision making where information is more deeply assessed against the goals of the landholder. Participation at this level of decision making is important in gaining understanding of the process of adoption and adaptation that determines the fate of conservation practices. The next section includes suggestions of how credibility and trust may be achieved.

The conduct of extension

Here we briefly present some suggestions about the conduct of extension that can be related to the findings we have reported earlier. It is not an exhaustive manual of extension methods, but focuses on core issues related to enhancing adoption.

Any sound extension campaign needs to use multiple methods (Vanclay 2004). Multiple extension channels, repetition, multiple deliverers of the message, and harnessing of peer pressure are among the standard tools of effective extension agents. Reliance on any particular method (e.g. print articles, verbal presentations, group extension, advertisements) will fall short of the potential impact on adoption from a diverse portfolio of extension approaches and channels. One advantage of using multiple approaches is that it increases the chances of reaching more of the relevant groups of landholders. In addition, different landholders have different learning styles and prefer to receive information in different ways, or through different channels (Bardsley 1982). Also, repetition can help to reinforce a message and build confidence, especially if it comes through different channels and from different sources.

Llewellyn *et al.* (2005) argued for a more sophisticated approach to selecting which landholder perceptions to target with extension. Through their research on farmer adoption of

Integrated Weed Management practices, they found that, of the many variables about which farmers held perceptions, only a small minority could be usefully targeted by extension. The majority of variables were either perceived accurately already (in which case there was nothing for extension to do) or would not be influential on adoption behaviour (in which case extension would not be useful). These insights appear to provide a useful pathway for improving the cost-effectiveness of extension.

In situations where adoption of innovations by a group of landholders confers reciprocal spillover benefits, developments in the theory of collective action (Ostrom 1998) suggest that adoption of these innovations may be enhanced by promoting the learning of social norms that emphasise mutual benefit/reciprocal benefits, and by efforts to build the mutual trust within that group (Marshall 2004b, 2005).

A notable trend in extension practice in Australia over the last 15 years has been the substantial decline in public funding for traditional one-on-one extension and a rise in group-based extension (Marsh and Pannell 2000). It could be argued that group-based extension has never been funded at a level that would allow its efficacy to be comprehensively tested (Curtis 2000). Nevertheless the data of Mullen *et al.* (2000) show that increased funding for what could broadly be called group-based environmental extension (including group facilitators) has roughly offset the decline in traditional production-oriented extension. Group-based extension is, of course, an important part of the extension system, but like any extension approach it has its limitations (Vanclay 2004). In the 1990s, group-based extension processes came to be relied on in the National Landcare Program, partly in response to perceptions about their ability to harness peer pressure to address what were often perceived (incorrectly in some cases) to be environmental problems requiring collective action by landholders for their effective resolution. Group-extension processes grew in favour among extension theorists in response to an increased emphasis on adult learning principles and participation by stakeholders (Chamala and Keith 1995; Knowles 1984; Röling 1988). They were embraced by state agriculture agencies, in significant part, for budgetary reasons (Barr 1994; Marsh and Pannell 2000).

Although group-extension approaches are undoubtedly useful, the swing from individuals to groups may have gone too far. For example, the introverted personality profiles of graziers described in the work by Shrapnel and Davie (2001) indicate the continued importance of one-on-one extension. Noting the importance of credibility in effective extension, Vanclay (2004, p. 221) observed that, 'Credibility is developed over time through the provision of credible, practical, useful answers that assist farmers in [their] day-to-day operations. Group facilitators who never provide on-farm advice rarely develop credibility and their ideas are easily dismissed'.

We have previously discussed the importance of credibility, trust and confidence in extension agents on the part of a landholder. A history of valuable advice relevant to a landholder's goals is probably the single most important source of credibility, but it can be enhanced to some extent by a wide range of factors, including: (1) authority and technical expertise of the extension agent; (2) perceived similarity of the extension agent to their audience; (3) local profile of the extension agent (e.g. local residence); (4) communication skills of the extension agent; (5) personal relationships between the extension agent and landholders; and (6) extension-agent acknowledgment of/empathy with the circumstances and problems of landholders.

Adviser credibility and trust is a valuable commodity, but it is only earned slowly. Adviser credibility and trust can be easily lost by the support of an innovation or practice clearly unsuited to local circumstances, or through the evangelical promotion of a practice that is clearly in conflict with the goals of landowners. In the past 20 years, the role of government extension agents in many states has changed away from that of supporting landholders in making good decisions to achieve their own goals, towards encouraging landholders to make decisions that achieve outcomes for the public good. In many situations, this has the potential to reshape the social contract between adviser and landholder, creating a far more complex social interaction that may be less comfortable for both. The importance of this changed social relationship is not recognised by the relevant public agencies, which publicise their programs using the rhetoric of community development, yet place clear requirements for technology transfer outcomes upon their agents.

Implications for policy and for environmental bodies

As noted in the introduction, some government officers express frustration at the lack of adoption by landholders of conservation practices and call for additional social research to understand adoption better. Sometimes it can be helpful to improve understanding of the adoption of specific practices, but the influences on adoption in general have been studied intensely and we believe that they are sufficiently well understood. Rather than more research into adoption, the more pressing need is to apply what is already well established in the adoption literature.

As we have seen, one implication is that if a practice is not adopted in the long term, it is because landholders are not convinced that it advances their goals sufficiently to outweigh its costs. A consequence of this is that we should avoid putting the main burden for promoting adoption onto communication, education and persuasion activities. This strategy is unfortunately common, but is destined to fail if the innovations being promoted are not sufficiently attractive to the target audience. The innovations need to be 'adoptable'. If they are not, then communication and education activities will simply confirm a landholder's decision not to adopt, as well as degrade the social standing of the field agents of the organisation. Extension providers should invest time and resources in attempting to ascertain whether an innovation is adoptable before proceeding with extension to promote its uptake.

For some environmental issues, the real challenge is to find or develop innovations that are not only good for the environment, but also economically superior to the practices they are supposed to replace. If such innovations cannot be identified or developed, there is no point in falling back onto communication. Promoting inferior practices will only lead to frustration for all parties.

Sometimes unattractive practices can be made sufficiently attractive by the provision of financial incentive payments (e.g. through economic policy instruments). However, it is important to be realistic about the potential of this approach. In some cases, the level of payment required to achieve sufficient adoption would be more than can be justified by the resulting environmental benefits (e.g. Pannell 2001a). In some situations, the most sensible strategy is *not* to attempt to encourage uptake of existing technologies or systems. Rather, it may be more sensible to attempt to develop better practices (more effective or more adoptable), or it may be that research and policy needs to address the task of living with the problem.

In conclusion, we set out to provide an integrated review of several disciplinary literatures on the adoption of conservation practices by rural landholders. We found that many of the findings and perspectives of our separate disciplines are consistent and readily translatable across disciplinary boundaries. We discussed these findings in three broad groupings: those relating to adoption as a process of learning, those relating to characteristics of potential adopters, and those relating to characteristics of the conservation practice. In general, adoption of conservation practices is complex and multifaceted, but it is, nevertheless, reasonably well studied and understood. In light of the literature, the disappointing levels of adoption of conservation practices that are often observed are readily explicable in terms of characteristics of the learning process, the potential adopters or the conservation practices. We have identified a number of important implications of the review for research, extension and policy.

Acknowledgements

This chapter is revised from Pannell DJ, Marshall GR, Barr N, Curtis A, Vanclay F and Wilkinson R (2006) Understanding and promoting adoption of conservation practices by rural landholders. *Australian Journal of Experimental Agriculture* **46**(11), 1407–1424.

The authors are grateful to anonymous reviewers and to Amabel Fulton and Sally Marsh for their detailed and insightful suggestions. Funders who have contributed directly or indirectly to the preparation of this review include Land and Water Australia, the Australian Research Council, Rural Industries Research and Development Corporation, Grains Research and Development Corporation, and the CRC for Plant-Based Management of Dryland Salinity.

References

Abadi Ghadim AK and Pannell DJ (1999) A conceptual framework of adoption of an agricultural innovation. *Agricultural Economics* **21**(2), 145–154.

Abadi Ghadim AK and Pannell DJ (2003) Risk attitudes and risk perceptions of crop producers in Western Australia. In *Risk Management and the Environment: Agriculture in Perspective.* (Eds BA Babcock, RW Fraser and JN Lekakis) pp. 113–133. Kluwer: Dordrecht.

Abadi Ghadim AK, Pannell DJ and Burton MP (2005) Risk, uncertainty and learning in adoption of a crop innovation. *Agricultural Economics* **33**(1), 1–9.

Anderson AM (1979) *How Advisors Advise: Agricultural Extension As a Social Process.* Hawkesbury Agricultural College: Richmond.

Anderson AM (1981) *Farmers' Expectations and Use of Agricultural Extension Services.* Hawkesbury Agricultural College: Richmond.

Australian Bureau of Agricultural and Resource Economics (2003) 'Australian farm surveys report 2002'. ABARE: Canberra.

Australian Bureau of Statistics (2002) 'Salinity on Australian farms'. Report 4615.0. Australian Bureau of Statistics: Canberra.

Bardsley B (1982) *Farmers' Assessment of Information and Its Sources.* School of Agriculture and Forestry, University of Melbourne: Melbourne.

Bardsley P and Harris M (1987) An approach to the econometric estimation of attitudes to risk in agriculture. *Australian Journal of Agricultural Economics* **31**(2), 112–126.

Barr N (1994) Landcare from inside-out and outside in. *Australian Farm Manager* **5**, 2–10.

Barr N (2004) 'The micro-dynamics of occupational and demographic change in Australian agriculture: 1976–2001'. Australian Bureau of Statistics: Canberra.

Barr N and Cary J (2000) *Influencing Improved Natural Resource Management on Farms: A Guide to Factors Influencing the Adoption of Sustainable Natural Resource Management Practices.* Bureau of Rural Sciences: Canberra.

Barr N and Wilkinson R (2005) Social persistence of plant-based management of dryland salinity. *Australian Journal of Experimental Agriculture* **45**(11), 1495–1501.

Barr NF (1999) Salinity control, water reform and structural adjustment: the Tragowel Plains irrigation district. PhD thesis. Institute of Land and Food, University of Melbourne: Melbourne.

Black A (2000) Extension theory and practice: a review. *Australian Journal of Experimental Agriculture* **40**(4), 493–502.

Bond G and Wonder B (1980) Risk attitudes amongst Australian farmers. *Australian Journal of Agricultural Economics* **24**(1), 16–34.

Cancian F (1979) *The Innovator's Situation*. Stanford University Press: Stanford.

Carlson GA and Wetzstein ME (1993) Firm decisions and behavior in pest management on a regional level. In *Agricultural and Environmental Resource Economics*. (Eds GA Carlson, D Zilberman and JA Miranowski) pp. 273–288. Oxford University Press: Oxford.

Cary J, Webb T and Barr N (2001) *The Adoption of Sustainable Practices: Some New Insights. An Analysis of Drivers and Constraints for the Adoption of Sustainable Practices Derived from Research*. Bureau of Rural Sciences: Canberra.

Cary J, Webb T and Barr N (2002) *Understanding Landholders' Capacity to Change to Sustainable Practices: Insights About Practice Adoption and Social Capacity for Change*. Bureau of Rural Sciences: Canberra.

Cary JW (1986) *Farmers' Attitudes to Land Management for Conservation*. School of Agriculture and Forestry, University of Melbourne: Melbourne.

Cary JW (1993) The nature of symbolic beliefs and environmental behaviour in a rural setting. *Environment and Behavior* **25**(4), 555–576.

Cary JW and Weston RE (1978) *Social Stress in Agriculture: The Implications of Rapid Economic Change*. School of Agriculture and Forestry, University of Melbourne: Melbourne.

Cary JW and Wilkinson RL (1997) Perceived profitability and farmers' conservation behaviour. *Journal of Agricultural Economics* **48**(1–3), 13–21.

Chamala S and Keith K (Eds) (1995) *Participative Approaches for Landcare*. Australian Academic Press: Brisbane.

Coughenour CM and Chamala S (2000) *Conservation Tillage and Cropping Innovation: Constructing the New Culture of Agriculture*. Iowa State University Press: Ames.

Curtis A (2000) Landcare: approaching the limits of volunteer action. *Australasian Journal of Environmental Management* **6**(1), 26–34.

Curtis A and Byron I (2002) *Understanding the Social Drivers of Catchment Management in the Wimmera*. Charles Sturt University: Albury.

Curtis A and Robertson A (2003) Understanding landholder management of river frontages: the Goulburn Broken. *Ecological Management and Restoration* **4**(1), 45–54.

Curtis A, MacKay J, Van Nouhuys M, Lockwood M, Byron I and Graham M (2000) 'Exploring landholder willingness and capacity to manage dryland salinity: the Goulburn Broken catchment'. Johnstone Centre report no. 138. Charles Sturt University: Albury.

Curtis AL (1997) Landcare, stewardship and biodiversity conservation. In *Frontiers in Ecology: Building the Links*. (Eds NI Klomp and I Lunt) pp. 143–153. Elsevier Science: Oxford.

Curtis AL and De Lacy T (1996) Landcare in Australia: does it make a difference? *Journal of Environmental Management* **46**(2), 119–137.

D'Emden FH, Llewellyn RS and Burton MP (2006) Adoption of conservation tillage in Australian cropping regions: an application of duration analysis. *Technological Forecasting and Social Change* **73**(6), 630–647.

Dong D and Saha A (1998) He came, he saw (and) he waited: an empirical analysis of inertia in technology adoption. *Applied Economics* **30**(7), 893–905.

Duncan RC (1969) An investigation of the 'trial' stage in the adoption process: aerial topdressing in the Clarence Valley. *Review of Marketing and Agricultural Economics* **37**(4), 207–216.

Feder G and Umali D (1993) The adoption of agricultural innovations: a review. *Technological Forecasting and Social Change* **43**(3–4), 215–239.

Feder G, Just R and Zilberman D (1985) Adoption of agricultural innovations in developing countries: a survey. *Economic Development and Cultural Change* **33**(2), 255–297.

Gasson R and Errington A (1993) *The Farm Family Business*. CAB International: Wallingford.

George RJ, Nulsen RA, Ferdowsian R and Raper GP (1999) Interactions between trees and groundwaters in recharge and discharge areas – a survey of Western Australian sites. *Agricultural Water Management* **39**(2–3), 91–113.

Geroski P (2000) Models of technology diffusion. *Research Policy* **29**(4–5), 603–625.

Gibbs M, Lindner R and Fischer AJ (1987) The discovery of innovations by farmers. *Journal of the Australian Institute of Agricultural Science* **53**(4), 254–261.

Goodwin BK and Schroeder TC (1994) Human capital, producer education programs, and adoption of forward-pricing methods. *American Journal of Agricultural Economics* **76**(4), 936–947.

Griliches Z (1957) Hybrid corn: an exploration in the economics of technological change. *Econometrica* **25**(4), 501–523.

Griliches Z (1960) Hybrid corn and economics of innovation. *Science* **132**(3422), 275–280.

Guerin LJ and Guerin TF (1994) Constraints to the adoption of innovations in agricultural research and environmental management: a review. *Australian Journal of Experimental Agriculture* **34**(4), 549–571.

Hagerstrand T (1967) *Innovation Diffusion As a Spatial Process*. The University of Chicago Press: Chicago.

Havens AE and Rogers EM (1961) Adoption of hybrid corn: profitability and the interaction effect. *Rural Sociology* **26**(4), 409–414.

Hawkins HS and Watson AS (1972) 'Shelford: a preliminary report of a social and economic study of a Victorian soldier settlement area'. Agricultural Extension Research Unit, University of Melbourne: Melbourne.

Helms GL, Bailey D and Glover TF (1987) Government programs and adoption of conservation tillage practices on nonirrigated wheat farms. *American Journal of Agricultural Economics* **69**(4), 786–795.

Howden P and Vanclay F (2000) Mythologisation of farming styles in Australian broadacre cropping. *Rural Sociology* **65**(2), 295–310.

Howden P, Vanclay F, Lemerle D and Kent J (1998) Working with the grain: farming styles amongst Australian broadacre croppers. *Rural Society* **8**(2), 109–125.

Janis M and Mann L (1977) *Decision Making: A Psychological Analysis of Conflict, Choice and Commitment*. Macmillan: New York.

Kaine G, Bewsell D, Boland A and Linehan C (2005) Using market research to understand the adoption of irrigation management strategies in the stone and pome fruit industry. *Australian Journal of Experimental Agriculture* **45**(9), 1181–1187.

Kaine GW and Lees JW (1994) 'Patterns in innovation: an analysis of the adoption of practices in beef cattle breeding'. Occasional paper no. 190. Rural Development Centre, University of New England: Armidale.

Kebede Y (1992) Risk behavior and new agricultural technologies: the case of producers in the central highlands of Ethiopia. *Quarterly Journal of International Agriculture* **31**, 269–284.

Kilpatrick S (2000) Education and training: impacts on farm management practice. *Journal of Agricultural Education and Extension* **7**(2), 105–116.

Kington EA and Pannell DJ (2003) Dryland salinity in the upper Kent River catchment of Western Australia: farmer perceptions and practices. *Australian Journal of Experimental Agriculture* **43**(1), 19–28.

Kingwell R, Hajkowicz S, Young J, Patton D, Trapnell L, Edward A, Krause M and Bathgate A (2003) 'Economic evaluation of salinity management options in cropping regions of Australia'. Report to the Grains Research and Development Corporation and the National Dryland Salinity Program: Canberra.

Kingwell RS, Pannell DJ and Robinson SD (1993) Tactical responses to seasonal conditions in whole-farm planning in Western Australia. *Agricultural Economics* **8**(3), 211–226.

Knowles M (1984) *The Adult Learner: A Neglected Species*. Gulf Publishing Company: Houston.

Latta J (2002) *A Survey of Farming Practices in the Low Rainfall Mallee Regions of New South Wales, South Australia and Victoria*. Department of Natural Resources and Environment: Walpeup.

Leathers HD and Smale M (1991) A Bayesian approach to explaining sequential adoption of components of a technological package. *American Journal of Agricultural Economics* **73**(3), 734–742.

Lindner RK (1987) Adoption and diffusion of technology: an overview. In *Technological Change in Postharvest Handling and Transportation of Grains in the Humid Tropics*. ACIAR Proceedings No. 19. (Eds BR Champ, E Highley and JV Remenyi) pp. 144–151. Australian Centre for International Agricultural Research: Canberra.

Lindner RK, Pardey PG and Jarrett FG (1982) Distance to information source and the time lag to early adoption of trace element fertilizers. *Australian Journal of Agricultural Economics* **26**(2), 98–113.

Llewellyn RS (2002) Adoption of integrated weed management by grain growers. PhD thesis. School of Agricultural and Resource Economics, University of Western Australia: Perth.

Llewellyn RS, Pannell DJ, Lindner RK and Powles SB (2005) Targeting key perceptions when planning and evaluating extension. *Australian Journal of Experimental Agriculture* **45**(12), 1627–1633.

Lockie S, Mead A, Vanclay F and Butler B (1995) Factors encouraging the adoption of more sustainable cropping systems in south-east Australia: profit, sustainability, risk and stability. *Journal of Sustainable Agriculture* **6**(1), 61–79.

Lodge GM (1991) Management practices and other factors contributing to the decline in persistence of grazed lucerne in temperate Australia: a review. *Australian Journal of Experimental Agriculture* **31**(5), 713–724.

Makeham JP and Malcolm LR (1993) *The Farming Game Now*. Cambridge University Press: Cambridge.

Marra M, Pannell DJ and Abadi Ghadim A (2003) The economics of risk, uncertainty and learning in the adoption of new agricultural technologies: where are we on the learning curve? *Agricultural Systems* **75**(2–3), 215–234.

Marsh S, Pannell D and Lindner R (2000) The impact of agricultural extension on adoption and diffusion of lupins as a new crop in Western Australia. *Australian Journal of Experimental Agriculture* **40**(4), 571–583.

Marsh SP (1998) 'What can agricultural researchers do to encourage the adoption of sustainable farming systems?'. SEA Working paper 98/05. Agricultural and Resource Economics, University of Western Australia: Perth.

Marsh SP and Pannell DJ (2000) Agricultural extension policy in Australia: the good, the bad and the misguided. *Australian Journal of Agricultural and Resource Economics* **44**(4), 605–627.

Marsh SP, Burton MP and Pannell DJ (2006) Understanding farmers' monitoring of water tables for salinity management. *Australian Journal of Experimental Agriculture* **46**(7), 1113–1122.

Marshall GR (2004a) From words to deeds: enforcing farmers' conservation cost-sharing commitments. *Journal of Rural Studies* **20**(2), 157–167.

Marshall GR (2004b) Farmers cooperating in the commons? A study of collective action in salinity management. *Ecological Economics* **51**(3–4), 271–286.

Marshall GR (2005) *Economics for Collaborative Environmental Management: Renegotiating the Commons*. Earthscan Publications: London.

Mesiti L and Vanclay F (2006) Specifying the farming styles in viticulture. *Australian Journal of Experimental Agriculture* **46**(4), 585–593.

Miranowski JA and Carlson GA (1993) Agricultural resource economics: an overview. In *Agricultural and Environmental Resource Economics*. (Eds GA Carlson, D Zilberman and JA Miranowski) pp. 3–27. Oxford University Press: New York.

Miranowski JA, Hrubovak J and Sutton J (1991) The effects of commodity programs on resource use. In *Commodity and Resource Policies in Agricultural Systems*. (Eds N Bockstael and R Just) pp. 275–292. Springer-Verlag: New York.

Mues C, Chapman L and Van Hilst R (1998) 'Survey of Landcare and land management practices: 1992–93'. Australian Bureau of Agricultural and Resource Economics: Canberra.

Mullen JD, Vernon D and Fishpool KI (2000) Agricultural extension policy in Australia: public funding and market failure. *Australian Journal of Agricultural and Resource Economics* **44**(4), 629–645.

Nicoll R (1994) *Landholders' Knowledge and Perceptions of Dryland Salinity Control Options in the Broadford and Kilmore Shires*. Environmental Management Research Project, Deakin University: Melbourne.

Ohlmer B, Olson K and Brehmer B (1998) Understanding farmers' decision making processes and improving managerial assistance. *Journal of Agricultural Economics* **18**, 273–290.

Ostrom E (1998) A behavioral approach to the rational choice theory of collective action. *American Political Science Review* **92**(1), 1–22.

Ostrom E, Gardner R and Walker J (1994) Regularities from the laboratory and possible explanations. In *Rules, Games, and Common-pool Resources*. (Eds E Ostrom, R Gardner and J Walker) pp. 195–220. University of Michigan Press: Ann Arbor.

Pannell DJ (1996) Lessons from a decade of whole-farm modelling in Western Australia. *Review of Agricultural Economics* **18**(3), 373–383.

Pannell DJ (1999) Social and economic challenges in the development of complex farming systems. *Agroforestry Systems* **45**(1–3), 395–409.

Pannell DJ (2001a) Dryland salinity: economic, scientific, social and policy dimensions. *The Australian Journal of Agricultural and Resource Economics* **45**(4), 517–546.

Pannell DJ (2001b) Explaining non-adoption of practices to prevent dryland salinity in Western Australia: implications for policy. In *Land Degradation*. (Ed. A Conacher) pp. 335–346. Kluwer: Dordrecht.

Pannell DJ, McFarlane DJ and Ferdowsian R (2001) Rethinking the externality issue for dryland salinity in Western Australia. *Australian Journal of Agricultural and Resource Economics* **45**(3), 459–475.

Phillips TI (1985) The development of methodologies for the determination and facilitation of learning for dairy farmers. Master's thesis. School of Agriculture and Forestry, University of Melbourne: Melbourne.

Presser HA and Cornish JB (1968) *Channels of Information and Farmers' Goals in Relation to the Adoption of Recommended Practices: A Survey in a Dried Fruit Growing District*. University of Melbourne, School of Agriculture: Melbourne.

Rahm MR and Huffman WE (1984) The adoption of reduced tillage: the role of human capital and other variables. *American Journal of Agricultural Economics* **66**(4), 405–412.

Ransom K and Barr N (1994) 'The adoption of dryland lucerne in northcentral Victoria'. Research Report Series No. 151. Department of Agriculture: Melbourne.

Reeve I (2001) 'Australian farmers' attitudes on rural environmental issues: 1991–2000'. Report to Land and Water Australia. Institute for Rural Futures, University of New England: Armidale.

Reeve I, Frost L, Musgrave W and Stayner R (2002) 'Agriculture and natural resource management in the Murray Darling Basin: policy history and analysis'. Report to the Murray Darling Basin Commission. Institute for Rural Futures, University of New England: Armidale.

Ridley AM and Pannell DJ (2005) The role of plants and plant-based R&D in managing dryland salinity in Australia. *Australian Journal of Experimental Agriculture* **45**(11), 1341–1355.

Rogers EM (1962) *Diffusion of Innovations*. Free Press: New York.

Rogers EM (2003) *Diffusion of Innovations*. 5th edn. Free Press: New York.

Röling NG (1988) *Extension Science, Information Systems in Agricultural Development*. Cambridge University Press: Cambridge.

Ruttan VW (1996) What happened to technology adoption–diffusion research? *Sociologia Ruralis* **36**(1), 51–73.

Shampine A (1998) Compensating for information externalities in technology diffusion models. *American Journal of Agricultural Economics* **80**(2), 337–346.

Shapiro BI, Brorsen BW and Doster DH (1992) Adoption of doublecropping soyabean and wheat. *Southern Journal of Agricultural Economics* **24**(2), 33–40.

Shrapnel M (2002) Bushies and cockies – beyond the myths: the personalities of our outback land managers. Master's thesis. University of Queensland: Brisbane.

Shrapnel M and Davie J (2001) The influence of personality in determining farmer responsiveness to risk. *Journal of Agricultural Education and Extension* **7**(3), 167–178.

Sinden JA and King DA (1990) Adoption of soil conservation measures in Manilla Shire, New South Wales. *Review of Marketing and Agricultural Economics* **58**(2–3), 179–192.

Sobels J, Curtis A and Lockie S (2001) The role of Landcare networks in rural Australia: exploring the contribution of social capital. *Journal of Rural Studies* **17**(3), 265–276.

Stoyles M (1992) *Cultural Barriers to Extension: Results of a Survey of the Non English Speaking Background Farmers in East Shepparton*. Ethnic Communications: Melbourne.

Tonks I (1983) Bayesian learning and the optimal investment decision of the firm. *Economic Journal* **93**(369), 87–98.

Tsur Y, Sternberg M and Hochman E (1990) Dynamic modelling of innovation process adoption with risk aversion and learning. *Oxford Economic Papers* **42**(2), 336–355.

Vanclay F (1986) Socio-economic correlates of adoption of soil conservation technology. Master's thesis. Department of Anthropology and Sociology, University of Queensland: Brisbane.

Vanclay F (1992) The social context of farmers' adoption of environmentally sound farming practices. In *Agriculture, Environment and Society: Contemporary Issues for Australia*. (Eds F Vanclay and B Furze) pp. 94–121. Macmillan: Melbourne.

Vanclay F (1997) The social basis of environmental management. In *Critical Landcare*. (Eds S Lockie and F Vanclay) pp. 9–27. Centre for Rural Social Research, Charles Sturt University: Wagga Wagga.

Vanclay F (2002) Conceptualising social impacts. *Environmental Impact Assessment Review* **22**(3), 183–211.

Vanclay F (2004) Social principles for agricultural extension to assist in the promotion of natural resource management. *Australian Journal of Experimental Agriculture* **44**(3), 213–222.

Vanclay F and Lawrence G (1995) *The Environmental Imperative*. Central Queensland University Press: Rockhampton.

Vanclay F and Lockie S (1993) *Barriers to the Adoption of Sustainable Crop Rotations*. Centre for Rural Social Research, Charles Sturt University: Wagga Wagga.

Vanclay F, Mesiti L and Howden P (1998) Styles of farming and farming subcultures: appropriate concepts for Australian rural sociology? *Rural Society* **8**(2), 85–107.

Vanclay F, Howden P, Mesiti L and Glyde S (2006) The social and intellectual construction of farming styles. *Sociologia Ruralis* **46**(1), 61–82.

Weston RE and Cary JW (1979) *A Change for the Better: Stress, Attitudes and Decision Making of Dairy Farmers 1976 to 1978*. School of Agriculture and Forestry, University of Melbourne: Melbourne.

Wiebers UC (1992) Economic and environmental effects of pest management information and pesticides: the case of processing tomatoes in California. PhD thesis. Technical University of Berlin: Berlin.

Wilkinson R and Cary JW (1992) *Monitoring Landcare in North Central Victoria*. School of Agriculture and Forestry, University of Melbourne: Melbourne.

Wilkinson RL (1989) Stepwise adoption of a complex agricultural technology. Master's thesis. University of Melbourne: Melbourne.

Wilkinson RL and Cary JW (1993) *Monitoring Soilcare in North East Victoria*. School of Agriculture and Forestry, University of Melbourne: Melbourne.

3

The many meanings of adoption

Roger Wilkinson

Summary

On the surface, adoption seems a simple concept. In various contexts, people express a desire for increased adoption of new technology by farmers without being clear about what they mean. In fact the concept of adoption is complex and its meaning is not clear. Adoption is not an event but a complex process with multiple dimensions. The aim of this chapter is to explore some of the complexity behind the term, 'adoption'. The main points are:

- Adoption is not a steady state but a continuous process;
- Adoption may be partial or incomplete in that a technology might be adopted to different extents on different farms or in different regions;
- Adoption may proceed gradually, through increasing extent or intensity of use of a new technology;
- A package of related technologies may be adopted in a stepwise manner;
- The same technology may be used in different ways on different properties;
- Technologies are not static but evolve and are adapted by users to fit different situations; and
- A technology may be disadopted at any time.

By explaining these points, I hope to encourage those people interested in increased adoption to be clear about what they are trying to achieve.

Introduction

It is easy to express a desire for increased adoption by farmers of a particular technology. But what does 'increased adoption' actually mean? How is it expressed in the actual practice of farmers? My aim in this chapter is to explain that the meaning of the seemingly simple word 'adoption' is not at all clear. Adoption is not an event; it is a process, and a complex process with multiple dimensions. By explaining some of these dimensions, I hope to encourage clarity of thought among the policy-making and extension communities about just what they are trying to achieve when they express a desire for increased adoption.

There is a need to distinguish between adoption and diffusion. To make this distinction, one must differentiate between technology uptake at the individual level (the process of discovery, decision and action through which each individual goes when taking up a new practice, generally called adoption) and at the aggregate social level ('the spread of a phenomenon over space and through time' [Gregory 2000, p. 175], generally called diffusion). Diffusion has almost always been conceptualised and measured more continuously than adoption, or at least using more categories, but it is important to think of adoption as something that is as continuous as diffusion.

Adoption as a continuous process

In the early days of adoption research, the decision whether or not to adopt a technology was generally treated as dichotomous. The adoption variable was usually categorised as binary: the farmer was classified either as an 'adopter' or a 'nonadopter' (Feder *et al.* 1985). There was little discussion about whether or not adoption could be partial or incomplete. The adoption decision was considered to be discrete and categorical. It was also assumed that the precise point of adoption could be specified exactly.

Even in many recent studies, adoption has often been measured as a binary decision. In their recent review of adoption research, Knowler and Bradshaw (2007) collated the findings of numerous studies since the mid-1980s: about half of them used a dichotomous measure of adoption. As a current example, Vignola *et al.* (2010), despite their complex jargon and air of sophistication, achieved a measure of adoption that was more than binary by simply counting the number of practices farmers had adopted.

For some technologies, this is probably all the precision that needs to be considered. Examples are television, tractors and rotary dairy sheds. These are 'lumpy' technologies, unable to be broken down; they have to be adopted either totally or not at all. Moreover, the evidence of their adoption is readily observable: one day the cows are milked in the new shed.

For many technologies, however, the dichotomous adoption assumption oversimplifies the adoption process. This is particularly true of technologies that are difficult to understand and implement, or are composed of several parts. Many adoption decisions are continuous, and there are two dimensions to this continuity. There is the process of finding out about the technology, deciding whether to adopt, then actually adopting (or not adopting), and there is the degree to which the new technology is used once it has been adopted.

Recognition that simple models do not explain adoption adequately has led to the presentation of a plethora of multi-stage models of the adoption process. That of Rogers (2003) is the most widely known, but there are several other multi-stage models, such as that of Sinden and King (1990). However, while multi-stage models might capture reality more effectively than simpler models (if only by virtue of being more comprehensive), it is still a guess as to when the point of 'adoption' actually occurs. This is because of the continuous nature of the adoption process. Rogers, perhaps the greatest exponent of stage models of adoption, recognised that evidence for distinct stages is tentative (Rogers 2003, p. 198). While it is generally agreed that there are different steps in the adoption process, not all steps are present in every instance of adoption, and sometimes different steps occur together and are inseparable. Occasionally they may occur in a different order from that which is expected from the model.

The precise point of 'adoption' is difficult to define, and most definitions appear circular. Tornatzky *et al.* (1983, p. 24) observed that the only consensus was in the definition of adoption as the central event in the process, 'the point which divides the organisation not having the technology from its having it'. There are several different ideas as to when adoption actually occurs: whether it is the point of making a mental commitment to use the technology, the point of initiating trial, or the time when full-scale commercial use is begun. Some have even questioned the value of the concept of adoption itself, claiming that the point of adoption is usually decided in retrospect by the 'weight of the evidence' (Tornatzky *et al.* 1983, p. 24). In the case of adoption that requires the purchase of new machinery, the point of purchase is obvious enough. But when a technology consists of a new way of using existing machinery or a slightly different practice, then the point of adoption probably depends on the purpose of the research and the level of analysis (Tornatzky *et al.* 1983, p. 24). At the very least (following Tornatzky *et al.* 1983, p. 25) there is a need to distinguish conceptually and methodologically between adoption as the entire process and adoption as a single event in a model of the process.

Incomplete or partial adoption

The early behaviourists assumed a link between knowledge and behaviour, and this assumption was carried forward in the diffusion model of technology adoption, which assumed that once someone had heard about some new technology (and was then persuaded of its advantages), they would then follow the rational course, which was to adopt it (Hooks *et al.* 1983, p. 309). Any difference in the rate of adoption between individuals was attributed to differences in individual 'innovativeness' (e.g. Gartrell and Gartrell 1979).

This assumption has been challenged in two ways, and both challenges occurred soon after the diffusion model was promulgated. Many years ago Presser (1969) recognised that innovativeness was contextual, that someone might adopt one technology early yet another technology late because the contexts in which the person might use the two technologies were different. A similar point has also been made more recently (Hoffmann 2007; Kaine 2004; Pannell *et al.* 2006). Devotees of the current trend in business and government circles for using the word 'innovation' to mean a state of being innovative – the review of Australia's 'innovation system' by Cutler (2008) is a good example – rather than a technology to be adopted, would do well to heed this point. Innovativeness is only part of the story: context is critical.

The other challenge to the 'innovativeness' assumption was made many years ago by Griliches (1957), who showed that a substantial proportion of the state-by-state variation in the rate of acceptance of hybrid corn in the United States was explicable by 'differences in the profitability of the shift to hybrids in different parts of the country' (Griliches 1957, p. 519). This observation was confirmed subsequently by Dixon (1980). The economic constraint model was based on the assumption that economic barriers prevent the direct link between knowledge and behaviour assumed in the diffusion model.

Central to Griliches' observation was the idea of locational specificity of new technology, reflected in the differential profitability of hybrid corn in different localities. Later, however, such locational specificity was seen as being due to environmental and ecological influences on adoption (Cary 1987). Rather than asking about the profitability of a technology, researchers started asking, for example, does the technology fit the farmer, does it fit the district? (Ashby 1982; Perrin and Winklemann 1976). However, ecologists and economists appear to have been saying the same thing from two different perspectives: that the advantages to be had from adoption of a new technology vary according to locality, as 'a more favourable environment increases the expected utility of net income' (Hiebert 1974, p. 767).

A central tenet of the ecological perspective is that farmers do not always adopt a technology completely (Feder *et al.* 1985). Griliches' (1957) assumption of incomplete adoption was shown by Dixon (1980) to be erroneous, and in hindsight the overwhelming yield advantage of hybrid corn over traditional open-pollinated varieties is obvious and clear. However, many complicated modern technologies may not exhibit such overwhelming advantages over traditional systems, or at least may have benefits which are more diffuse than those of hybrid corn. Thus it is important to realise that farmers might not adopt a technology completely. In this situation, the fact that a farmer has adopted a new variety is imprecise information; it is not known whether the farmer has sown it on a trial plot, one paddock or the whole farm. The level of adoption of a technology should not be determined by a simple average of the number of adopters over the total population of potential adopters. A more accurate representation is an aggregate of the extent of adoption by the individuals in the social system under consideration.

Gradual adoption

Adoption may be not only partial but also gradual. Gradual adoption can be defined as increasing area or intensity of use over time of a single technology. It reflects the possibility of partial adoption by permitting termination of the gradual adoption process at any point and, as it implies a process, it allows a mechanism for the achievement of partial or incomplete adoption. The notion of gradual adoption implies that one does not simply arrive at a state of adoption, but must go through a process of gradually increasing adoption to reach it. Examples are a gradually increasing area sown to a new seed variety and a gradually increasing rate of application of a new fertiliser.

Ryan (1948, p. 275) recognised that, with hybrid corn, 'it is not enough to deal simply with the fact of "acceptance". Hybrid seed may first be accepted for trial on just a few acres, or it may be adopted on a 100 per cent basis'. This notion of divisibility for trialing has been used in numerous studies, but its main use has been to explain differing rates of adoption between innovations. For example, Rogers (2003, p. 258) generalised: 'The trialability of an innovation, as perceived by the members of a social system, is positively related to its rate of adoption'. The concept of 'trialability' has been used to compare innovations, but not to follow their trials.

Of course, gradual adoption has some similarities to Rogers' concept of trialability, but it differs in other ways, and goes beyond trialability in still others. Gradually increasing the area sown to a new variety might well also be part of a trial of the variety but increasing the application rate of a new seed, fertiliser or herbicide is not. The former is a method of uncertainty reduction, which is what Rogers (2003, p. 258) intended for his concept. The latter may have some role in providing information by demonstrating the appropriate production function, but its purpose is not the trial of an innovation to determine its suitability for adoption. The reason for trial is lack of information; gradual adoption exists for other reasons. For example, the farmer might be constrained by lack of working capital or might initially underestimate the optimum input application rate.

Ryan and Gross (1943), in their study of diffusion of hybrid corn, observed a pattern of increasing acceptance of the new hybrids over time; each operator sowing only a part of his corn acreage to hybrid corn in his first year of adoption, then increasing the proportion in subsequent years. And in his later, broader study, Ryan (1948, p. 280) noted, 'There was probably for each operator a gradual season-to-season increase in the percentage of crop planted in hybrid seed'. While gradual adoption was recognised in both studies, in neither was a distinction between trial and gradual adoption noted.

Duncan (1969) provided such a distinction. He asked, 'Is thirty acres of top-dressed pasture a trial or is it as much as the landholder can, or will, do for economic, social, or other reasons?' (Duncan 1969, p. 208). He felt that the difference between trial and adoption lay in the farmer's actions subsequent to their first attempt at using the new technology. 'For instance, if a farmer, after treating a very small area of the farm with superphosphate for a number of years, considerably increased the area fertilised, it could be fairly concluded that such action was more likely to indicate a trial stage than if the farmer had gradually increased the area fertilised' (Duncan 1969, p. 213). Moreover, he noted the appearance of a continuous progression over time as use of the practice of aerial topdressing increased, thus realising that adoption can be as continuous as diffusion.

Carlson and Dillman (1988) were able to distinguish between trial users of no-till (a method of cropping without cultivation) and those who had adopted the practice, adding a further dimension to Duncan's distinction between trial and gradual adoption. Their separation was based on the farmer's degree of commitment to no-till as a farming practice. Commitment was measured by a combination of how many times a farmer had used no-till in the past and how likely the farmer was to use it in the future. They were thus able to separate the farmers in their survey into two groups: those who had used no-till in the past but had not adopted it and those who had adopted it. While Carlson and Dillman did not explicitly mention the possibility of gradual adoption, they did recognise that the point of adoption was difficult to define, because adoption was not instantaneous. Such a concern is similar to that of Tornatzky *et al.* (1983, p. 24) mentioned earlier.

The distinction between trialing and gradual adoption is confounded by the reduced necessity for extensive trialling, as a new technology is adopted by more and more people (Barr and Cary 2000). Presser (1969, p. 511) observed, 'once the idea is well tested and proved to be useful in an area, and individuals adopt it without intensive experimentation, it is a practice rather than an innovation'. In Duncan's (1969) study, the earliest farmers to trial the new technology started with small areas, while the later adopters trialled larger areas and progressed to full-scale adoption more quickly. Ewers (1988, pp. 52–54) noticed the same phenomenon in the adoption of laser-controlled grading: later adopters trialled larger and larger areas, until eventually it was impossible for an observer to distinguish between a trial and full-scale adoption.

Stepwise adoption

Some technologies are packages of separate component technologies, each of which could be adopted separately. The classic technology package in agriculture is the green revolution in less-industrialised countries, consisting of elements such as high-yielding crop varieties, chemical fertilisers, new planting methods and chemical weed control.

The components of a technological package may be easy or difficult to disentangle, perhaps depending on how tightly the components are bundled together. Rogers (2003, p. 186) described a 'tightly bundled innovation' as 'a collection of highly interdependent components', such that 'it is difficult to adopt one element without adopting the other elements'. In contrast, 'a loosely bundled innovation consists of elements that are not highly interrelated; such an innovation can be flexibly suited by adopters to their conditions'. Lionberger (1960, p. 24) recognised as early as 1960 that, 'Complex practices that can be divided into much smaller parts and taken a little at a time are likely to be accepted more quickly than where this is not possible'.

The components of a technological package are often adopted in a stepwise manner. Byerlee and Hesse de Polanco (1986) showed that, by exploiting critical complementarities, Mexican farmers adopted technological packages in steps of only one or two components are a time. Wilkinson (1989) showed similar stepwise adoption patterns for permanent beds, a complex technological package for irrigation cropping in northern Victoria. Wilkinson (1989) also showed that farmers generally do not adopt all components of a complex technological package, instead preferring to partially adopt it by using only those components that they perceive to fit their farming system.

Flexibility in use

Some technologies are particularly flexible and can be used in different ways. A computer, for example, can be used in many different ways, with different software, for different purposes, by different kinds of users. An agricultural technology may well need to be fully integrated into the farm production system if the farmer is to gain full advantage of it. If the technology is sufficiently flexible that it may be used in different ways, it may be able to be adopted to some extent without being fully integrated into the farm production system. It may, for example, be able to be used in a particular niche on the farm, perhaps to gain one particular benefit that may be highly desired by the farmer, rather than fully integrating it into the farming system. In this way, the farmer can capture some of the benefits of the technology without making radical changes to the farm system.

The flexibility to use a technology in different ways can increase the likelihood of adoption, because it gives farmers choices. In recent research conducted with mixed farmers in Western Australia, lucerne (*Medicago sativa*) has been identified as a complex but flexible technology (Wilkinson and Dolling 2008). Lucerne can be used in a niche capacity for watertable management without making major changes to mixed farming systems. However, some farmers recognised additional benefits of growing lucerne, such as weed control and additional out-of-season feed, but to achieve the full suite of benefits of lucerne they had to make the major changes necessary to integrate lucerne into their farm systems. Some farmers began with lucerne as niche users and integrated it into their farm system at a later time. Other farmers began by integrating it but later became niche users.

Technological evolution

A technology is not constant, but keeps changing even as it is being adopted. For researchers used to being in control of 'their' technology and to treating its users as 'respectful waiting publics' (Gieryn 1999, p. x), this can be a difficult concept to grasp. Nowak (1984, p. 225) observed that, 'It appears researchers have the working assumption that once a technology is used, it remains constant thereafter. Researchers also seemed to assume that the technologies emerging from their experimental plots remained constant through the diffusion process'.

Technologies change and evolve over time as improved versions are developed, refinements are made, and feedback is given from early users (Clark 1986, pp. 79–80). Both researchers and users are involved in efforts to improve the technology. In fact, as Nowak (1984, p. 225) noted in relation to reduced tillage technologies, 'some of the most innovative and promising research is taking place on farmers' fields'. Carlson and Dillman (1986, p. 91) went further, observing that no-till was a farmer innovation, and urging that farmers and scientists must communicate to ensure the continued evolution of the technology. More recently, particularly within the science studies literature, it has been recognised that a technology continues to be made and remade, both by scientists and farmers, through refinement, adaptation, interrogation and maintenance, until it fits the current context for which it is required (Carr and Wilkinson 2005).

Adaptation

Not only does a technology change while it is being extended and diffused, but it is likely to be adopted in different versions by different people. This adaptation was described by Nowak (1984, p. 225) as 'the attempt by the operator to increase the agronomic and economic efficiency of the adopted technology'. While researchers may feel they have a 'proprietary right' to a new technology, or at least a 'psychological involvement' (Larsen and Agarwala-Rogers 1977), adaptation will probably promote greater adoption of the technology, by increasing the potential advantage of the new technology over the old. In this context, adaptation can be seen as a functional response to overcome the locational specificity of a technology. The farmers who were doing the 'promising research' noted by Nowak (1984, p. 255) were doing it because they perceived the reduced tillage technology being introduced to them as inappropriate to their farm and incompatible with their existing farming system. They were simply trying to make it work on their farm.

I prefer Nowak's term 'adaptation' to the term 'reinvention' used by Rogers (1983) and others because it implies a modification of an existing technology to make it more useful in a different area, rather than a complete rearrangement of the technology at the whim of the 'reinventor'. 'Reinvention' often implies change that is more extensive than does 'adaptation' (Tornatzky *et al.* 1983, p. 136). Sometimes, it implies change for unnecessary reasons, as in 'reinventing the wheel'. 'Adaptation' implies change for more constructive reasons, and it implies that the need for alteration is more real than does 'modification'. Adaptations are not made solely for their own sake. Rogers (2003, p. 184) recently conceded that reinvention is not necessarily bad, but has maintained his use of the term.

Some technologies appear to require or encourage adaptation more than others. Larsen and Agarwala-Rogers (1977) found that the more difficult an idea was to implement, the less likely it was to be adopted unchanged. As Rice and Rogers (1980, p. 501) observed, 'problems with operational implementation of an innovation may be one impetus for reinvention: The innovation as originally introduced to a system may not match with that system's problem'. Carlson and Dillman (1988) found that the greater a farmer's mechanical skill, the more likely the farmer was to have adopted no-till technology. They argued that farmers with more mechanical skill were better able to perform the adaptation of the technology required to make it fit their own unique situation. This was seen to be more important for a system-type technology (such as no-till) than for a less complex technology that could be adopted without changing other aspects of the farming system. Adaptation appears to be necessary for the successful extension and adoption of new cropping systems.

Without adaptation, adoption is likely to be slow and, in many cases, inappropriate. Recognising the existence of adaptation, it would perhaps be better to help farmers adapt the new technology to suit their own farm, rather than expecting them to adopt what someone else considers to be the 'best available system'. Adaptation (and thus adoption) is clearly aided by a technology that provides flexibility and a range of options.

Disadoption

The last dimension of continuity in the adoption process that I want to discuss is the idea that adoption is not the end of the process. Even after adoption has occurred, the technology may be disadopted for a variety of reasons. While Rogers recognised the existence of 'discontinuance' in 1962 (pp. 88–95), in many cases he considered it to be almost pathological, often the result of not using the new technology properly. Later, however, he elevated 'discontinuance' to the status of something understandable, sometimes desirable, in certain circumstances (Rogers 1983, pp. 186–191). Others have been more explicit. Swanson *et al.* (1986, p. 111) warned that farmers may reject new soil conservation technologies even after they have adopted them because of high maintenance requirements or incompatibility with other new technologies. Pannell *et al.* (2006) noted that disadoption may occur because economic circumstances (e.g. changes in market prices) reduce the relative advantage of a technology.

Another way technologies can be disadopted is when they are superseded. Brennan and Cullis (1987) modelled the replacement of one wheat variety by another, based on attributes of the different varieties. They noted that the disadoption phase can be as important as the adoption phase. Goldenberg and Oreg (2007) took this line further, arguing that where a product progresses through generations (their example was portable audio moving from cassettes, through compact discs and minidiscs, to MP3 players), someone with laggardly tendencies may skip a generation of the product and become an early adopter of a later product. In making this argument, they recast the laggard, made famous by Rogers (1962), from someone who is slow to adopt the new into someone who is slow to discadopt the old.

When one technology replaces another, the progress in adoption of the new technology is not always a mirror image of the progress in disadopting the old. Mattingly (1987), a geographer, assessed the patterns of adoption of tractors and disadoption of horses among the rural counties of Illinois, USA. He found that disadoption of horses lagged behind adoption of tractors, but the lag was shorter for later adopting counties than for earlier adopting ones. In a county where adoption of tractors occurred relatively early, 50% of farms had tractors by about 1932, yet 50% still had horses in 1950, a lag of 18 years. In contrast, in a late adopting county, the 50% figure for both tractors and horses occurred around 1954. This phenomenon is similar to that observed earlier for trialing, where the adoption process is accelerated for later adopters.

For more complicated technologies, Kislev and Shchori-Bachrach (1973) documented an 'innovation cycle' whereby an industry undergoes a constant process of adoption of new technology and disadoption of obsolete technology, with producers possessing different skills and business structures adopting at different stages in the cycle. Adoption cannot be seen as the final stage in the process, because users continually evaluate both the new technology and their use of it.

Conclusion

I have tried to explain some of the ways in which the seemingly simple word 'adoption' represents not an event or a steady state, but a continuous process. Partial adoption occurs when a technology is adopted to different extents on different farms or in different regions. Gradual adoption occurs when a farmer increases the extent of use of a new technology over the farm, or increases the intensity of use of the technology. Stepwise adoption occurs where related technologies are aggregated together and promoted as a package, yet farmers adopt the component technologies in steps rather than all at once. Where a technology is sufficiently flexible, different farmers may use it in different ways.

All these elements of continuous adoption are further complicated by the propensity of both farmers and scientists to adapt new technologies to better suit different circumstances and the fact that a technology may be disadopted at any point (mostly, but not always, in favour of a newer technology). Further, there are yet more dimensions of the continuity of the adoption process that I have not described, such as the decision-making process a farmer undertakes when considering whether to adopt a technology, and the diffusion of a new technology through a social system. Although my examples have been mainly agricultural technologies, my points apply equally to the adoption of new technologies in other fields of endeavour.

Some of the points I have made are not new. Indeed, more than 40 years ago Presser (1969, p. 512) said, 'Use of a new idea or the criterion for adoption also needs specification. Adoption may be defined as continued full scale use since the first trial; increasing scale of use since first trial; trial, use, and later discontinuance; trial and then discontinuance; or just trial alone'. However, little has been written on the continuous nature of adoption over the past 20 years, and I have had to quote mostly old references. There is still a need to learn from the writings of earlier researchers.

In all the ways I have described, adoption is a complex and continuous process. Rather than simply calling for increased adoption, policy makers and extension managers and practitioners would do well to consider just what aspects of adoption they would like increased. The word 'adoption' is so entrenched in the language that everyone who uses the word thinks they know what is meant by it, so changing the way in which the word is used is not an easy thing to do. Because it is so hard to know what is meant by the term 'adoption', I think it may well be the wrong term to use. It would be better to focus not on farmers' **adoption** of technology but on their **use** of technology.

References

Ashby JA (1982) Technology and ecology: implications for innovation research in peasant agriculture. *Rural Sociology* **47**(1), 234–250.

Barr NF and Cary JW (2000) *Influencing Improved Natural Resource Management on Farms: A Guide to Factors Influencing the Adoption of Sustainable Natural Resource Management Practices.* Bureau of Resource Sciences: Canberra.

Brennan JP and Cullis BR (1987) 'Estimating the adoption and disadoption of wheat cultivars'. Paper presented to the 31st annual conference of the Australian Agricultural Economics Society. University of Adelaide: Adelaide.

Byerlee D and Hesse de Polanco E (1986) Farmers' stepwise adoption of technological packages: evidence from the Mexican altiplano. *American Journal of Agricultural Economics* **68**(3), 519–527.

Carlson JE and Dillman DA (1986) Early adopters and nonusers of no-till in the Pacific Northwest: a comparison. In *Conserving Soil.* (Eds SB Lovejoy and TL Napier) pp. 83–92. Soil Conservation Society of America: Ankeny, Iowa.

Carlson JE and Dillman DA (1988) The influence of farmers' mechanical skill in the development and adoption of a new agricultural practice. *Rural Sociology* **53**(2), 235–245.

Carr AJL and Wilkinson RL (2005) Beyond participation: boundary organizations as a new space for farmers and scientists to interact. *Society and Natural Resources* **18**(3), 255–265.

Cary JW (1987) 'Current problems and barriers to information transfer of irrigation technology'. Paper presented to the Murray Darling Basin Ministerial Council workshop on the transfer of irrigation technology. Yanco, New South Wales.

Clark G (1986) Diffusion of agricultural innovations. In *Progress in Agricultural Geography.* (Ed. M Pacione) pp. 70–92. Croom Helm: Beckenham, Kent.

Cutler T (2008) *Venturous Australia: Building Strength in Innovation, A Review of the National Innovation System.* Department of Innovation, Industry, Science and Research: Canberra.

Dixon R (1980) Hybrid corn revisited. *Econometrica* **48**(6), 1451–1461.

Duncan RC (1969) An investigation of the 'trial' stage in the adoption process: aerial topdressing in the Clarence valley. *Review of Marketing and Agricultural Economics* **37**(4), 207–216.

Ewers CR (1988) 'Laser grading in the Goulburn-Murray irrigation district: an innovation diffusion study'. Monash Publications in Geography No. 35. School of Geography, Monash University: Clayton, Victoria.

Feder G, Just RE and Zilberman D (1985) Adoption of agricultural innovations in developing countries: a survey. *Economic Development and Cultural Change* **33**(2), 255–298.

Gartrell JW and Gartrell CD (1979) Status, knowledge and innovation. *Rural Sociology* **44**(1), 73–94.

Gieryn TF (1999) *Cultural Boundaries of Science: Credibility on the Line.* University of Chicago Press: Chicago.

Goldenberg J and Oreg S (2007) Laggards in disguise: resistance to adopt and the leapfrogging effect. *Technological Forecasting and Social Change* **74**(8), 1272–1281.

Gregory D (2000) Diffusion. In *The Dictionary of Human Geography*. 4th edn. (Ed. RJ Johnston) pp. 175–178. Blackwell: Oxford.

Griliches Z (1957) Hybrid corn: an exploration in the economics of technological change. *Econometrica* **25**(4), 501–522.

Hiebert LD (1974) Risk, learning, and the adoption of fertilizer responsive seed varieties. *American Journal of Agricultural Economics* **56**(4), 764–768.

Hoffmann V (2007) Book review: five editions (1962–2003) of Everett Rogers's Diffusion of Innovations. *Journal of Agricultural Education and Extension* **13**(1), 147–158.

Hooks GM, Napier TL and Carter MV (1983) Correlates of adoption behavior: the case of farm technologies. *Rural Sociology* **48**(2), 308–323.

Kaine GW (2004) 'Consumer behaviour as a theory of innovation adoption in agriculture'. Social Research Working Paper No. 01/04. AgResearch: Hamilton, New Zealand.

Kislev Y and Shchori-Bachrach N (1973) The process of an innovation cycle. *American Journal of Agricultural Economics* **55**(1), 28–37.

Knowler D and Bradshaw B (2007) Farmers' adoption of conservation agriculture: a review and synthesis of recent research. *Food Policy* **32**(1), 25–48.

Larsen JK and Agarwala-Rogers R (1977) Re-invention of innovative ideas: Modified? Adopted? None of the above? *Evaluation* **4**, 136–140.

Lionberger HF (1960) *Adoption of New Ideas and Practices*. Iowa State University Press: Ames.

Mattingly PF (1987) Patterns of horse devolution and tractor diffusion in Illinois, 1920–1982. *Professional Geographer* **39**, 298–309.

Nowak PJ (1984) Adoption and diffusion of soil and water conservation practices. In *Future Agricultural Technology and Resource Conservation*. (Eds BC English, JA Maetzold, BR Holding and EO Heady) pp. 214–237. Iowa State University Press: Ames.

Pannell DJ, Marshall GR, Barr NF, Curtis A, Vanclay F and Wilkinson RL (2006) Understanding and promoting adoption of conservation practices by rural landholders. *Australian Journal of Experimental Agriculture* **46**(11), 1407–1424.

Perrin R and Winklemann D (1976) Impediments to technical progress on small versus large farms. *Agricultural Economics* **58**, 888–894.

Presser HA (1969) Measuring innovativeness rather than adoption. *Rural Sociology* **34**(4), 510–527.

Rice RE and Rogers EM (1980) Reinvention in the innovation process. *Knowledge: Creation, Diffusion, Utilization* **1**, 499–514.

Rogers EM (1962) *Diffusion of Innovations*. Free Press of Glencoe: New York.

Rogers EM (1983) *Diffusion of Innovations*. 3rd edn. Free Press: New York.

Rogers EM (2003) *Diffusion of Innovations*. 5th edn. Free Press: New York.

Ryan B (1948) A study in technological diffusion. *Rural Sociology* **13**, 273–285.

Ryan B and Gross NC (1943) The diffusion of hybrid seed corn in two Iowa communities. *Rural Sociology* **8**(1), 15–24.

Sinden JA and King DA (1990) Adoption of soil conservation measures in Manilla Shire, New South Wales. *Review of Marketing and Agricultural Economics* **58**(2–3), 179–192.

Swanson LE, Camboni SM and Napier TL (1986) Barriers to adoption of soil conservation practices on farms. In *Conserving Soil*. (Eds SB Lovejoy and TL Napier) pp. 108–120. Soil Conservation Society of America: Ankeny, Iowa.

Tornatzky LG, Eveland JD, Boylan MG, Hetzner WA, Johnson EC, Roitman D and Schneider J (1983) *The Process of Technological Innovation: Reviewing the Literature*. National Science Foundation: Washington D.C.

Vignola R, Koellner T, Scholz RW and McDaniels TL (2010) Decision-making by farmers regarding ecosystem services: factors affecting soil conservation efforts in Costa Rica. *Land Use Policy* **27**(4), 1132–1142.

Wilkinson RL (1989) Stepwise adoption of a complex agricultural technology. Master of Agricultural Science thesis. University of Melbourne: Melbourne.

Wilkinson RL and Dolling PJ (2008) 'What happens when a complex innovation meets an existing production system? Case studies of lucerne growers in Western Australia'. Paper presented to the Primary Industries Innovation Symposium. University of New England: Armidale.

4

Social principles for agricultural extension in facilitating the adoption of new practices

Frank Vanclay

Summary

An understanding of the social context of agriculture, the social nature of farming, and the socio-cultural and socio-psychological basis of adoption is needed if agricultural extension is to be effective in facilitating the adoption of new practices and enabling change. Some 27 principles are presented, with the key points being to acknowledge that farming is a social activity, to recognise the diversity of farmers, to appreciate that adoption is a socio-cultural process as well as a social-psychological process and not simply a process of communication, and to reflect on what all this means for extension practice. To be effective in enabling change, extension staff need to consider adoption and extension as social change and social learning processes.

Introduction

Agriculture has too long been thought of as a technical activity involving the application of science and the linear transference of the products of that science via a top-down process of technology transfer. It is not. Agriculture is farming, and farming is people. The future of agriculture is dependent on the survival of healthy, viable and vital rural communities that have resilience and dynamic capacity to respond to a wide range of issues. This chapter seeks to outline the key social principles relevant to enabling change in individuals, family farm businesses, communities and industries involved with primary industries and natural resource management (NRM).

The principles were developed out of a personal reflection on now over 25 years of research into the social dimensions of farming. This research program started with a study of the socio-economic correlates of adoption of soil conservation technology on the Darling Downs as part of a Masters degree at the University of Queensland (Vanclay 1986) and continued with a PhD at Wageningen University in The Netherlands (Vanclay 1994) which looked at the wider socio-political context of agriculture and extension in Australia. It involved many consultancies and academic research projects, as well as the supervision of many PhD students across a range of topics in rural social research, especially relating to the concept of 'styles of farming' (Howden *et al.* 1998; Howden and Vanclay 2000; Mesiti and Vanclay 2006; Vanclay *et al.* 1998, 2006,

2007). It draws on work undertaken in conjunction with colleagues including Roy Rickson (Rickson *et al.* 1987), John Cary (Vanclay and Cary 1989), Stewart Lockie (Lockie *et al.* 1995; Vanclay and Lockie 1993) and others, for a range of organisations including NSW Agriculture, the Cooperative Research Centre for Viticulture, and the Cooperative Research Centre for Weed Management Systems. Over the years, the ideas have developed and consolidated, benefiting greatly from being aired at many conferences and in being published in various forms (notably Vanclay 1992a, 1997, 2004a; Vanclay and Lawrence 1995).

They are called 'principles' because they are intended to be regarded as 'a general law or doctrine that is used as a basis of reasoning or a guide to action or behaviour' (Australian Oxford Paperback Dictionary 1989). While this status may not be accorded to them by all agricultural scientists, they do have that status from a rural social research perspective. It is the argument of this chapter that agricultural scientists and extension staff should accept these statements as principles.

Principles

Principle 1: Farming is a socio-cultural practice

The first principle is to acknowledge that farming is a socio-cultural practice rather than just a technical activity. More than just a job, farming is a way of life, a way of making a living, and has social meaning much deeper than most other occupational identities. In that sense, farming is a vocation, and has a deep sense of place (Vanclay 2008). As a socio-cultural practice, it is governed, informed and regulated by social processes. Being aware of this fact, and thinking reflectively about what this understanding means will assist in the promotion of a sustainable agriculture for Australia's future.

Principle 2: Farmers are not all the same

The farming community is not homogeneous. There are many ways in which diversity can be observed within the farming community: rich and poor, big and small, old and young, early in the life cycle or late in the life cycle, high mortgage and small mortgage, propensity to adopt new ideas (innovator) and propensity to retain tried and true methods ('laggard' in extension discourse), pro-chemical (or pro-GMO) and anti-chemical (or anti-GMO), and male and female. Farmers (family farm businesses) can be categorised on every single variable that can be logically considered in conjunction with agriculture. This means there are no single problems, no single solutions, no single extension strategies, and no best medium that extension should solely utilise.

Instead of classifying family farm businesses according to demographic or structural variables as has been undertaken by extension researchers in the past (e.g. adopter vs. non-adopter, innovator vs. laggard, big vs. small, old vs. young, valley floor vs. hillside), it may be more meaningful to group farmers into subcultural groupings representing a conglomerate of social and structural variables, which can be called 'styles of farming' (Howden *et al.* 1998; Vanclay *et al.* 1998, 2006, 2007). The farming styles concept is an heuristic that allows for an understanding of the range of worldviews about how to farm. Appreciating the existence of a range of worldviews is important. Different family farm businesses have different priorities, different understandings, different values, different ways of working, and different problems. Extension should address the needs of all styles, and not just those that align with the techno-scientific orientation often promoted by agricultural science and extension agencies.

Principle 3: Adoption is a socio-cultural process

Rather than extension being a process of communication between science as the only originator of ideas and farmers as passive adopters, extension needs to appreciate that adoption is a social process. The act of adoption is not an uncritical response to information provided by extension; it is a learning process by individuals and family farm businesses that considers a wide range of issues. Adoption is not a singular decision act of an individual in an isolated context. Adoption takes place in a social context, with farmers discussing their ideas with other members of the family farm business and with farmers. Much adoption occurs when the idea or practice to be adopted has become part of the normative concept of 'good farm management'. Adoption is a socio-cultural process, not an isolated decision.

Principle 4: Profit is not the main driving force of family farm businesses

Contrary to the expectations of many economists, extensionists and agricultural scientists, maximising profit is not the most important thing in the lives of most family farm businesses (Vanclay 1992a). Farmers seek to make a reasonable income for a reasonable amount of work taking a reasonable amount of risk, with each family farm business defining what is reasonable for itself. The additional values and virtues of being a farmer, i.e. the lifestyle factors, compensate farmers for those times when income may be less that what may be achieved by other endeavours. Appeal to economic incentives alone is not sufficient to bring about change.

Principle 5: It is hard to be green when you are in the red

Although profit is not the main driving force for most family farm businesses, and the promotion of adoption of new practices requires more than just appeal to the economic dimension, it is a truism that 'it is hard to be green when you are in the red'. There are two aspects to this saying. One relates to the need for the practices being promoted to at least work and be beneficial, and not be too costly. The second aspect relates to the difficulties family farm businesses face when they are financially and/or personally stressed as a result of an economic downturn or a drought. In periods of hardship (being in the red), farmers have simply no resources to expend, no matter how favourable a practice or an innovation might be. 'Resources' refers not only to financial capital and time, but also relates to the capacity to respond. One consequence of stress is a reduction in the perceived options available and an increased reluctance to take risks, a concept known as agricultural involution (Geertz 1963; Vanclay 2003). Thus capacity building is an important aspect of extension and the process of enabling change (SELN 2006).

Principle 6: 'Doing the right thing' is a strong motivational driver

Farmers do what they consider to be the 'right thing' and they conform to notions of 'good farm management' (Phillips and Gray 1995; Vanclay and Lawrence 1995). This notion is a complex entity and includes ideas about farming practice, environmental management, and also social matters about being a farmer and living in a rural community. There is also a gender dimension to this (Silvasti 2003; Vanclay *et al.* 2007). Good farm management is not a singular absolute. Rather, it is a dynamic concept that takes into account a family farm business's unique situation – their land, soils, debt situation and goals. Thus different farmers have varying notions of what good farm management actually implies. It is a social concept that is socially-constructed with individual farmers determining for themselves what constitutes good farm management through discursive interaction with other farmers, with extension officers (public and private), through reading farming literature and through exposure to the general media; in short, through participating in a farming subculture. Farmers' desires to conform to good farm management are responsible for much adoption of land-use practices, especially those that may not have economic advantages. Thinking about how to link recommended practices to the notion of good farm management may enhance their likelihood of adoption.

Principle 7: Farmers don't distinguish NRM issues from other farm management issues

Over the years, extension services to farmers provided by government and non-government organisations have tended to be split between production and NRM (or soil conservation) issues. One government department for production (DPI or agriculture) and another department for NRM (soil conservation service, conservation and land management, etc.). Even where there was one combined department, they tended to have separate divisions for production and NRM. This schism persists today with many non-government organisations or NRM regional bodies promoting NRM issues, but few thinking about production, or how NRM and production are intertwined. This division has led to different sets of extension staff promoting different messages, which were often contradictory. From a farmer's perspective, such a divide is ridiculous. From their perspective, there is only one farm. Farming practices have both production and NRM outcomes. Extension advice must therefore integrate NRM and production issues. Whole Farm Planning, Integrated Catchment Management, and the use of Environmental Management Systems in farming can, to some extent, assist in taking a holistic approach (Carruthers and Vanclay 2007), and should be more widely incorporated into extension activities.

Principle 8: There is a strong desire to hand the farm on to one's children

Most farmers want to pass the farm on to their children in a better condition than they themselves received it. This motivation is so strong that it exceeds any rational economic decision about the level of investment of labour, effort, money in improving the farm. A major problem occurs when farmers believe their children will not return to the farm because the motivation for investment in improvement is diminished. The problem is compounded because it is difficult to know precisely when or if children will return to the farm because sometimes children who have professional jobs in the city and have said all their adult lives that they will never return to farm will actually do so when their ageing parents pass away or announce that the farm is to be sold. At other times, even when children make it clear that they are not interested in farming, and it may be evident to all that it is highly unlikely that they will return to the farm, the parents may still harbour a secret belief that the children will return and make plans consistent with that false belief.

Parents' desires to have children remain on, or return to the farm are powerful expectations that can cause strong feelings of obligation in farming families, especially when the farm has been in the family for generations. These feelings of obligation mean that there may be very strong feelings for keeping the farm, against all economic reason. To give up the farm, or worse, to lose the farm, are often perceived to be a sign of personal failure. These feelings of expectation and obligation can cause problems in intergenerational transfer, especially since many farms cannot support two families, at least not at the level of most people's expectations about the amount of disposable income they should have. It has been suggested that succession issues are responsible for much rural suicide. Succession planning is probably inadequately undertaken by most farming families and needs to be the focus of increased attention by agricultural organisations.

Principle 9: Sustainability means staying on the farm

Many agencies want to increase sustainability in agriculture, but they tend to regard sustainability more in biophysical terms (the environment), and sometimes in economic terms. For farmers, the social significance of farming means that the social dimension of sustainability is central; sustainability is meaningless unless it involves their ability to stay on the farm. For farmers, therefore, sustainability means something along the lines of 'we as a family, on our farm, in the future'. The physical environmental dimension of sustainability is important, but a continued ability to make a living is more important. Looking after the land, stewardship, and now 'caring for our country', was always part of the notion of good farm management, and so for farmers sustainability is not a new concept even if the physical expression of this (in terms of what management practices should be used) has changed.

Principle 10: Not 'farm' and 'farmer', but 'family farm business' (women are an integral part of the farm)

A farm is rarely the embodiment of a singular individual male farmer. The word 'farmer' is a linguistic convenience that has an established romanticised meaning that belies the reality of farm management. Farms are often complex partnerships involving many people in the financial affairs and in the running of the farm and farm household. The gender blindness, if not sexism, that afflicts extension and agricultural science results in the role of women in agriculture being understated if not unrecognised. In the present and in the past, women play a major role in farm management, albeit one that is under-appreciated. Women's roles in agriculture are possibly increasing (or at least increasingly being recognised) and the number of female operators is growing now and will continue to do so. Even in individual situations where there has been a strong division of labour, the role of women in the private sphere in the household has been essential to the survival of the farm. Extension needs to acknowledge the role of women on the farm and needs to consider how the needs of women can be met, and how they can be engaged more (see other chapters in this book by Fulton and Vanclay, and Cathy McGowan).

Principle 11: Stage in the life cycle is significant

The stage in the life cycle of a farm family affects their need for household and disposable income, and this potentially affects finance available for other purposes. But stage in the life cycle also affects commitment to the future, with young families being more committed to a future on the farm than either families later in the life cycle, or young single farmers. Stage in the life cycle is therefore a complicated variable, but it demonstrates that there are many factors that are involved in decisions about adopting new management practices or technologies, and that adoption is not a simple process of communication.

Principle 12: 'Non-adoption' is not the cause of land degradation

Many extension staff believe that it is the non-adoption of the practices they promote that is the main barrier to achieving sustainable agriculture, and therefore express dismay about those farmers who do not adopt whatever practice is being promoted. However, an argument could be made that it is the *adoption* (not non-adoption) of many practices that were actively promoted in the past that is largely responsible for many of the environmental problems today. Examples supporting this include tree-clearing (which in many cases was a requirement of tenure) which has led to salinisation in many locations, and the excessive use of 'sub and super' (i.e. subterranean clover and super-phosphate) which was encouraged by financial incentives and which has been responsible for much soil acidification.

There are many farmers who feel that in the past extension has 'oversold' the benefits of the practices being promoted and that they feel vindicated in their decision not to adopt. Many farmers have witnessed the recommended practices (whether environmental or commercial) changing over time. As farmers are often much older than the extension officers they encounter, they are often surprised and unimpressed by the zealousness of some extension officers in suggesting 'if you only do such-and-such, all your problems will be over'. Many farmers have heard that every 10 years or so over their life. Farmers are therefore naturally and rightly sceptical about many extension messages.

Principle 13: Marginal farmers are not necessarily marginal because of their lack of management skills

There are many reasons why family farm businesses may be 'marginal'. Being marginal in a fiscal sense does not necessarily mean being small in land holding. Some farms are engaged in the production of high value commodities and only need a limited area to farm. Others may have better soils, higher rainfall or greater access to irrigation. Other families may primarily have lifestyle objectives.

Many agricultural enterprises are marginal in the sense of not producing returns on investment (total asset value) of 6 or 8% or more as would normally be expected in business. Farms that survive and prosper tend to be ones that manage their cash flow adequately, and that return enough income to live on, irrespective of the total value of the farm. Debt levels are an important part of cash flow and thus farm survivability, especially when interest rates are high. Farms that survive are often those that have low mortgages/overdrafts, and not necessarily those that are farmed more effectively in an agronomic sense. Thus farming families who inherit their farm debt-free may prosper irrespective of their agronomic skills, while other families may struggle with a debt burden no matter how good their skills are.

There are many factors that affect the survival of a farm. In Australia, there is a substantial amount of luck relating to rainfall (when and where), and around decisions regarding planting, harvesting times and decisions relating to sale price and futures contracts. While farmers can (and should) improve their management skills to reduce the risks of farming, many of the risks cannot be controlled and therefore there will always be a chance element. Other factors are important too. For example, off-farm work to augment farm income has long been important in assisting young farm families getting started and helping families manage in difficult periods. Thus families that have access to sources of off-farm income (through convenient location, available time by one or more members of the family, and availability of jobs) are strategically advantaged. It should be noted that there are structural and chance elements to this source of income. Unfortunately, there is a tendency for extension staff to regard in a disparaging way those farmers they consider to be marginal, even when the apparent marginality may be completely outside the control of those farmers.

Principle 14: Farmers' attitudes are not the problem

It is often thought that improving farm management requires extension and education programs to change farmers' attitudes. But ample survey research reveals that farmers' attitudes *per se* are not antagonistic to the environment (Rickson *et al.* 1987; Vanclay 1992a; Vanclay and Lawrence 1995). Farmers do not believe that they are 'raping the earth' while driving their tractors. Research has shown that they have positive attitudes about environmental management generally but they may have different views about what environmental management actually means, about how to implement it, and they may have concerns about whether the land management practices being promoted as sustainable are, in fact, sustainable and/or profitable. To some extent, this is intuitively obvious. It is not likely that farmers would have environmentally hostile views. The case of land clearing, for example, while being seen by some as environmental destruction can be understood from the perspective of many farmers as being 'land improvement'. Thus the problem is not one of farmers having the wrong attitude, but one of a difference of views about the right way to manage the farm, and about what constitutes 'good farm management'.

Principle 15: Farmers construct their own knowledge

It is inappropriate to believe that only 'Science' (as a social institution) can create knowledge that is then transferred to the public via extension. All individuals construct meaning (knowledge) about their own experiences of the world. The information that is transmitted to farmers via extension or the media is assessed against other information, knowledge and beliefs held. Nothing is accepted without evaluation. As the community becomes more empowered and more sceptical, 'authoritative' information is increasingly rejected. Science does not automatically have credibility and legitimacy, especially when the ideas being put forward contradict people's common sense understandings and everyday experience (Fleming and Vanclay 2010).

Farmers create their own knowledge through experimentation and trial, and through their own theorising. The knowledge of Science (i.e. the knowledge created by scientists) is used by farmers when it is consistent with their own understanding. Even then it is adapted to fit their own worldview (rather than adopted), and so the adoption process itself represents a form of scientific inquiry undertaken by farmers (i.e. 'science' as a methodology). The knowledge of Science is rejected when it is inconsistent with a farmer's worldview. Thus, farmers are their own scientists, theorising, hypothesising and experimenting. Sometimes the knowledge farmers create is especially adapted to local conditions. The utilisation of this local knowledge can improve the applicability of scientific knowledge (see Vanclay *et al.* 2009). Farmers also develop considerable knowledge about their own farm. They know the local history and local conditions and they use that information in their decision making and management. Within viticulture, for example, it was found that while many viticultural management systems required careful examination of grapevines for pests and diseases, and extension agencies promoted specific 'scouting' strategies, the precision expected in such scouting was rarely undertaken by grape growers. Instead of thorough examination of the whole vineyard, many farmers used their knowledge of local 'hot-spots', i.e. locations in the vineyard where pest and disease outbreaks were likely to occur first, to minimise their scouting effort (Glyde and Vanclay 1996). While it is appropriate to accept that farmers have local knowledge, it is important not to romanticise or overstate the applicability of that knowledge. Local knowledge is unlikely to provide immediate answers to new problems. Of course farmers do experiment and they may over time develop solutions to new problems, and this may help science and other farmers to overcome these problems. But farmers could develop partial solutions that treat the symptom but not the cause, and which could exacerbate the problem (or other problems) over time.

Principle 16: Effective extension requires more than the transfer of technology, it needs an understanding of the worldviews of farmers

Historically, extension was predicated on the notion that knowledge transfer was uni-directional; that science was the only originator of new ideas, information and knowledge; and that farmers were passive, grateful recipients of new technology and knowledge. It also held that all new ideas, if successfully extended, would be adopted. Non-adoption could only mean that information transfer had not taken place (i.e. inadequate communication) or there was a barrier to adoption (i.e. some reason why farmers could not adopt the new technology, such as a lack of money). This argument is somewhat absurd. Surely if it really did make sense for a farmer to adopt a new practice and a commitment to that innovation existed (that there really was a comparative advantage), a way would be found to adopt it no matter what difficulties there might be. Where there is no adoption, obviously a real commitment to that innovation does not exist (at least not yet) and non-adoption may well be a sensible strategy. There are lots of reasons why farmers may not have a real commitment to new technologies, and thus non-adoption may well be rational from the perspective of the farmer. Extension needs to be relevant to the needs of individual farmers, and needs to put their needs ahead of institutional priorities if it is to be successful.

Principle 17: Farmers have legitimate reasons for non-adoption

The reasons given by farmers for not adopting new techniques can be categorised into about 12 legitimate reasons for non-adoption (developed further from Vanclay 1992b).

(1) *Too complex.* In general terms, the more complex the innovation, the greater the resistance to adoption. Complexity makes the innovation more difficult to understand and generally requires greater management skills. This increases the risk associated with the innovation. Many environmental management practices are complex and require, amongst other things, a detailed understanding of physical processes. In some cases, farmers may comprehend what is being advocated to address a particular issue such as salinity; however, they may simply not believe or agree with the scientific explanation. Farmers are acting quite rationally by preferring to adopt less complex innovations over more complex ones and by not adopting complex practices at all.

(2) *Not easily divisible into manageable parts.* Divisibility allows for partial adoption. Farmers can adopt that part of an innovation they like or that is consistent with other farming objectives. Obviously, the more divisible into component parts an innovation is, the more likely it is to be adopted. Under the traditional model of adoption of commercial innovations, partial adoption is thought to inevitably lead to complete adoption. Partial adoption is viewed as a form of trial adoption. Where innovations are not divisible, they are not likely to be adopted, especially if they have other detracting attributes. In this case, farmers must be totally committed to the new innovation before adoption. Such a commitment is unlikely for a range of reasons, and consequently farmers are acting rationally when they do not adopt technologies that are not divisible.

(3) *Not compatible with farm and personal objectives.* Farmers are more likely to adopt innovations that are compatible with other farm and personal objectives. Where innovations are complex and indivisible, they are also likely to require major changes in the management of the farm and may not be compatible with other operations on the farm. Farmers' personal needs for the use of capital and income, such as for the education of children, expenditure on household goods, as well as farm requirements such as the purchase of new machinery, may mean that capital expenditure is not consistent with farm and personal goals at that point in time. The desire to maintain flexibility because of uncertainty means that innovations that do not encourage flexibility are likely to be resisted. Because of the fundamental changes to agricultural practices required to address many NRM issues, most environmental innovations are not compatible with current farm management practices. Non-adoption under these circumstances is rational from the farmer's point of view.

(4) *Not flexible enough.* Many new management practices reduce farmers' flexibility. Farmers like flexibility because it means that they can change commodities in response to market and climatic conditions. Perennial pastures lock farmers into grazing. Zero-tillage systems, with chemical control of weeds, restrict the range of crops that can be grown and the rotations of those crops. Farmers are quite likely to resist the adoption of new technologies that restrict their flexibility. With fluctuating market prices, farmers are acting rationally by wanting to maintain flexibility.

(5) *Not profitable.* Not all farm management practices that are offered to farmers are profitable, at least not in the perception of each farmer. Even where farmers accept that some

new practice might be profitable for some farmers, they will often argue that the conditions on their farm are different and they will find other reasons why they would be unlikely to achieve the same results. Furthermore, farmers know that it takes a few seasons to iron out all the bugs in a farming system and to achieve the maximum benefit, so there may be a few years of lowered income. Because of the economic situation of many farmers, they simply cannot afford such down-time, and it makes more sense for them to continue with a tried-and-true system from which they are confident that they can get a secure return, than to invite the uncertainty of change, even if it will result in improved profit in the longer term. Some innovations, such as sustainable cropping rotations, do not necessarily return profit in every year, but are alleged to increase profit in a gross margin analysis over the whole rotation. Potentially this requires farmers to forego profit (and absolute income) in some years of a rotation in the promise that profit will be increased in other years. But farmers, or more specifically their banks, have requirements of a cash flow in every year. Most environmental innovations rarely provide direct economic benefit to an individual farmer even if they are of benefit to the wider community, especially when future discounting techniques are applied (Quiggin 1987), yet they are still adopted, at least to some extent. If farmers based their adoption decision solely on economic criteria, there would be very little adoption of environmental innovations. Fortunately, farmers employ a range of criteria in their decision-making processes, and do what they consider to be the right thing to do (refer to Principle 6) as much as it is practicable. Nevertheless, it is a truism that the more expensive NRM practices are (in terms of immediate capital and labour outlay and cost:benefit ratio over time), the less likely adoption will be. If farmers were being strictly rational, little adoption of environmental innovations would occur. They ought not be criticised for not adopting when the economic situation does not warrant it. There is a certain irony in that farmers are criticised for not adopting practices that extension believes to be profitable, but they are also criticised for not adopting environmental innovations which are not profitable.

(6) *Capital outlay is too high.* In addition to the economics of the innovation in terms of whether or not it will increase profit, it is necessary also to consider the capital required to adopt the new technology. Much innovation requires considerable capital outlay in the form of new machinery, seeds, agrichemicals or earthworks. Often, adoption of new techniques may require the farmer to forego some income until the new system is established. In this situation, the farmer must have the resources not only to adopt the new technology, but also to survive the period until the new innovation produces income. In periods when farmers have negative incomes, declining land values and equity levels, many farmers have no borrowing power (despite the fact that interest rates are at relatively low levels). In other words, most farmers do not have the capital resources available to them to adopt any new technology that requires a substantial capital outlay. It should be noted that most banks regard farm investment as high risk and charge high risk margins, meaning that farmers may be paying 5 to 10% more for their farm loans than the average private owner-occupied housing loan. Even in periods of low interest rates, the rate for farm borrowings may still be higher than the return on capital invested on the farm. This means that it is economically irrational for farmers to borrow money – or even to be a farmer at all! In addition to the lack of capital to outlay, the farm financial crisis means that most farmers are unwilling to take large risks because failure might have severe consequences. Risk taking is more likely when the farmer can afford the consequences of failure.

(7) *Too much additional learning is required.* In addition to the financial costs associated with the adoption of new practices, there are often intellectual costs or learning effort.

Farmers may have to learn new ways of doing things. Many new recommended practices require greater knowledge about cropping systems and chemicals use, for example. This classification is similar to 'complexity', but relates to the knowledge base of the individual farmer rather than to an objective measure of complexity. Farmers are not unique in attempting to limit the amount of knowledge needed in order to conduct their operations; this is a common human characteristic.

(8) *Risk and uncertainty is too great.* Risk is a concept usually associated with commercial innovations and refers to farmers' concerns that the capital and other resources invested in adopting the new practices will result in higher costs or lower benefits than anticipated. Risk also applies to environmental innovations in that farmers need to be sure that the NRM practice being considered will actually provide the anticipated environmental outcomes. Farmers could expend resources adopting new technology, buying new machinery, or altering the management of their farm in order to be more sustainable only to find that the new practice fails to solve the environmental problems it was intended to solve. In this sense, the risk may be greater for environmental innovations than for commercial innovations. With commercial innovations the main risk is capital outlay and perhaps the yield of one season. With environmental innovations the risk includes the capital resources expended (often considerable when production strategies are required to be altered) and the production for that season as well as the production for future seasons if the environmental degradation is not stopped. While farmers do not necessarily make conscious and sophisticated analyses of the degrees of risk in adopting technology (the information required to do this is seldom available), they are aware of the implications of particular choices. The economic situation faced by farmers tends to promote an aversion towards risk and uncertainty.

(9) *There is conflicting information.* No new practice, especially when designed for NRM outcomes, is free of debate about its applicability or effectiveness. Farmers receive information from numerous sources which often contradict each other. In a situation where there is already some uncertainty, conflicting information further suggests that non-adoption is an appropriate management strategy.

(10) *Do not see that there is a problem (lack of appreciation of the problem).* Considerable research has established that farmers are likely to adopt NRM practices when, among other things, they consider themselves to be personally at risk from environmental degradation (see for example Rickson *et al.* 1987; Vanclay 1992a). However, much of the extension and NRM literature, and especially the general media, depict land degradation in its most dramatic forms: deep erosion gullies, salt encrusted pans, or exposed tree roots resulting from wind erosion. The presentation of land degradation in this dramatic form is counterproductive (Vanclay 1992a). While farmers may be made aware of the issue, they do not see the same degree of degradation occurring on their own farm and consequently believe they do not personally have a problem. They will claim this even when it is known that the problem may be serious in their own locality. Where farmers experience land degradation in such a severe form, they may feel powerless to address the problem, and adopt a fatalistic attitude rather than undertake any reclamation action or fundamentally change their management practices. Greater depiction of the early warning signs would be more effective in promoting awareness amongst farmers.

(11) *Lacking the physical infrastructure.* Agricultural commodity production requires certain physical infrastructure, such as handling facilities to enable the crop to be marketed. Historically that infrastructure was provided by government in the form of commodity marketing boards and other organisations which together provided a network of silos and railways as well as extension services to provide advice on issues. The existence of this

infrastructure meant that it was effectively impossible for farmers to grow anything that was not compatible with that infrastructure. More recently there has been greater flexibility created especially with the increased use of trucking of produce. Nevertheless, current concerns by governments to increase production of higher-value crops and the perception about the reluctance of farmers to grow new crops should be tempered by consideration of the history of agricultural production and the extent to which farmers are locked in by existing infrastructure.

(12) *Lacking the social infrastructure.* In the same way that a physical infrastructure exists as a mechanism to encourage production of particular crops and to inhibit others, a social infrastructure also exists. This social infrastructure refers to the social networks or social capital of farmers which provides a bank of knowledge and experience for farmers to utilise (Kilpatrick and Vanclay 2005). The accumulated knowledge of other farmers is usually regarded as a more valuable source of information than extension services. Except for a few maverick farmers who will give anything a go, most individual farmers do not want to be the only one doing a particular activity because they would have no social support to discuss any problems that might arise. There is safety in numbers.

Principle 18: Top-down extension is inappropriate

Vanclay and Lawrence (1995) identified five major criticisms of traditional top-down extension. While contemporary extension agencies are moving away from traditional extension practices, the ideology that supported top-down extension persists in subtle ways. It is worth reiterating those criticisms to help ensure that those problems are not manifested in modern extension. First, extension has uncritically accepted the products of agro-industrial agriscience and agribusiness, and has seen its task as simply being to unquestioningly promote those products, irrespective of their applicability to farmers or society. Second, this uncritical acceptance of these products and their adoption by farmers has led to considerable social and ecological impacts. Third, the adoption-diffusion model underpinning most traditional extension practice is premised on commercial innovation in which it is perceived that farmers would benefit. Thus it does not cater for environmental innovations, which may not be of economic benefit to individual farmers. Fourth, farmers' local knowledge has been marginalised, trivialised, subordinated and ignored. Finally, extension has utilised a psychological model of individual decision making and has ignored the social, political, cultural and historical context of agriculture and adoption behaviour.

Principle 19: The 80–20 rule is a self-serving delusion

There is a story or 'script' (Vanclay *et al.* 2007) that the top 20% of farmers produce 80% of the agricultural wealth. This story is used to justify the focusing of extension services on those 20% of farmers. Sometimes in a further attempt to legitimate this strategy it is argued that these top 20% of farmers provide role models for the remaining farmers and that the ideas extended to these farmers will trickle down or diffuse throughout the whole farming population. In this way, the work of extension officers is undertaken even while they sleep! Even when the trickle-down concept was not necessarily applied, the view had strong political support because it was felt that the bottom farmers were recalcitrant laggards who would not change, and who would not in any case be part of the future of Australian agriculture, partly because it was thought that they could not survive and would be structurally adjusted out of farming.

This view is a self-serving fantasy that is socially inequitable and dangerous from an NRM perspective. It is inequitable because it has legitimised extension to focus on the needs of the top farmers ignoring the needs of other farmers which may very well be different. Thinking about the farming styles concept (refer to Principle 2), it is not necessarily true that the 80%

would not adopt new ideas; it may just be that the practices being promoted and the manner in which they were promoted only fitted with the worldviews and learning styles of some of the styles of farming (types of farmers). Had there been attention given specifically to the needs of a greater range of farmers, perhaps the rate of adoption would have been greater.

The story is damaging to the environment in that many NRM issues affect all of the Australian landscape. The landscape scale and severity of issues such as salinity mean that we cannot be complacent in appealing to change in the practices in only a small percentage of farmers. While the potential salinity threat is not evenly distributed over the landscape, the farmers likely to be affected by salinity are not necessarily in the top 20% of farmers. NRM issues mean that we need to be concerned about the farming practices of all farmers.

Principle 20: Science and extension do not have automatic legitimacy or credibility

Many decades ago, Australian farmers placed a high degree of trust in the agricultural research and extension system; at least this is what was widely believed and accepted. Extension officers felt important delivering useful information to an eager and receptive farming population. Those days have gone, if they ever existed. Today, farmers are sceptical and dubious about the stated claims of practices being promoted. The high credibility the research institutions allegedly once had has been lost, and farmers no longer immediately accept what is being promoted as being factual. Extension in its various new forms needs to win back the trust of farmers.

Principle 21: Representation is not participation

As a general rule, participation is a good thing. The involvement of farmers on boards and committees is desirable, although there is a danger that such representation is simply tokenistic. The main criticism of representation is that it does not necessarily mean participation, certainly not of the full range of farming styles. A major concern is that the farmer representatives are seldom representative of all farmers. Often they are chosen not because they are farmers *per se*, but because they are farmers who have considerable experience appropriate to the business activities of the board or committee on which they seek to serve. Because of their corporate experience, their worldview and life circumstances are very different from most other farmers. Thus, only certain styles of farming are represented. Furthermore, because their worldview is so different, those corporate farmer representatives are unlikely to be able to speak on behalf of styles other than their own, except on matters that are common to all (or at least most) farmers.

Even where representatives are recruited from a broad range of farmers, such farmer representatives are seldom in the majority on any committee, and thus can easily be marginalised. This marginalisation is even greater for those farmer representatives who are not used to the formal meeting discourse. Farming, although an activity requiring considerable decision making, is not an area calling for abstract conceptualisation and articulation of ideas in the same way as expected in the formal committee discourse. Thus the formal discourse itself acts to subordinate farmers. The bind is that those farmers who become comfortable with the discourse become 'bureaucratised', accepting of the hegemonic agenda, and thus fail to represent farmers at all. Enabling participation and deliberation is more difficult than getting a few representatives on a committee.

Principle 22: The use of dramatic images is counter-productive

Vanclay (1992a) argued that because of the predominance of dramatic images in the general media and in extension literature, farmers' overall concern about degradation has become inflated (that is they have increased awareness) but unfortunately they do not perceive themselves to be at risk because the land degradation they experience is not as severe as the images being depicted. Vanclay and Cary (1989) identified that one of the issues in relation to adoption of salinity control methods was the lack of knowledge of the early warning signs, the salt indicator species. However, the problem with many early warning signs is that they are not specific to a single issue, and can easily be attributed to other reasons. For example, a poor germination rate, reduced prolificness, or reduced prevalence of a particular species in a pasture could be attributed to a lack of moisture, too much rain, hot weather, cold weather, poor quality seed, pests, diseases, or many other things. Sometimes the telltale signs of a problem become so common that they are simply disregarded; for example, few farmers regard muddy dams or dirty creek water as evidence of soil erosion. It is desirable that farmers develop an understanding of the land, and that they consider the NRM processes that may contribute to the features in the landscape they observe. This has come to be known as 'land literacy'. Farmers (and others) need to be able to read the land for what it is telling us about its health and about the health of our society and our production systems.

Principle 23: Put degradation into perspective

There are many technical definitions of 'land degradation'. However, what extension officers and scientists regard as degradation is not necessarily perceived as degradation by all farmers. Generally, this discord is perceived by extension as the failure of farmers to develop sufficient 'awareness' of the issue. But strictly speaking, degradation is a value judgement made about the acceptability of a rate of some process. Land degradation occurs because of naturally occurring geomorphological processes, albeit increased by human activity. To some extent, our fertile farming lands are the result of the same processes that are now regarded as 'degradation', only having occurred at a slower rate and over a much longer time period. Farm management practices accelerate the rates of these naturally occurring processes, with some practices causing them to occur at rapid rates and other practices being responsible for slower rates of change. Since these processes are naturally occurring, they occur irrespective of the farming practices used, and even if the decision is not to farm. Thus, the understanding of these processes as induced degradation, rather than as a natural process, represents a social understanding about the acceptability of the rate of the process. What rate of these processes is acceptable?

Nutrient decline and acidification (at rates believed to be a problem) are virtually inevitable outcomes of all farming activities because of the harvesting of crops and consequent removal of plant material. Soil structure decline and erosion are possibly also inevitable because of machinery use. Nutrient decline and acidification potentially can be corrected artificially through the application of fertilisers and lime respectively, although this is not sustainable in the long run. The socio-economic issue here is that the cost of rectification may exceed either or both the increased yield to be gained from rectification, and/or the cost of replacing the land with new land (buying out the neighbour). Farmers are aware of this. Thus, awareness of land degradation occurring on the farm does not mean that it is economically rational for farmers to take ameliorative action.

Economists (e.g. Quiggin 1987) have established that most ameliorative actions to prevent land degradation are not economically rational, especially when future discounting is applied and the discount rate (or interest rate) is high. Fortunately, farmers (and other people for that matter) are not economically rational beings. While farmers cannot be expected to do things that are manifestly not economical, the argument put repeatedly here is that economics alone does not determine farming practice.

It could be considered that if degradation is the loss of productive farm land, then the greatest form of degradation is not salinity or acidity, but the conversion of farm land to non-farm use, which usually occurs for urban expansion or rural residential development, including hobby farms and the whole sea-change/tree-change phenomenon. The impacts of this for Australian farmers are not so much in terms of lost land (which affects Australia as a nation, but doesn't affect farmers directly), but more in terms of increasing the price of land in those areas subject to this form of development to be beyond the reach of farming. This means that smaller farmers cannot expand. From a sustainable agriculture point of view, we should be concerned about protecting (zoning) our productive farmlands to protect them from conversion to non-farm use. Whatever attractions rural residential (urban fringe) blocks may have for those people who desire them, they are undesirable from a sustainability point of view. The issue of rural residential blocks causes many disputes between non-farming rural residents and farmers, particularly over issues like pest and weed control, chemical use, odours, noise and bushfire prevention. These are complicated issues, and there is potential for creative solutions, although they do not appear to be applied in many cases. It does give a different perspective, however, on the question of what is 'land degradation', and demonstrates the importance of a social analysis in answering that question.

In terms of other environmental issues, notably water use, farmers are not the only water users. Wasted water in industry and in domestic applications also reduces the water available for environmental flows. Farmers feel that urban users should make a contribution as well.

Principle 24: The best method of extension is multiple methods

One of the more frequent questions raised in extension discourse is: *'What is the best method of extension?'* The answer usually expected is a singular and simplistic response: facts sheets or farm visits or field days, etc. When the diversity of farmers is appreciated, and the socio-cultural basis of farming is understood, there can be only one answer; there is no singular best method of extension; multiple methods of extension are required to deliver the message to the diverse range of farmers, and to reinforce the message in different ways.

Principle 25: Group extension is not a panacea

With reduced public expenditures and a concern about private benefits, state governments are reducing publicly-funded extension services. However, there is still a recognised need for dissemination of an extension message, especially in relation to public good issues like NRM. Group extension is seen as an efficient way of communicating that message. Group extension does have many virtues, but it is not a solution to every issue. Ultimately, each farm is different and farmers use awareness of the differences of their farm as a way of justifying why a certain practice may not be appropriate to them. Individual, one-on-one extension is needed to assist with on-farm issues.

Extension is also a process where the credibility of the person giving the advice is an important factor in the weighting that farmers assign to that advice. Credibility is developed over time through the provision of credible, practical, useful answers that assist farmers in their day-to-day operations. Group facilitators who do not provide on-farm advice rarely develop credibility and their ideas are easily dismissed. Thus, a strong argument can be mounted that group extension also requires supporting one-to-one extension, and that the credibility of extension officers in a group setting is enhanced by their one-to-one extension experience.

Principle 26: Extension is likely to only have a small impact

This social understanding of farming and the adoption process creates the realisation that effecting extensive change (large changes and changes to a large percentage of the farming population) is unlikely. This does not mean that extension is ineffective or unsuccessful. It just means that there needs to be realistic expectations about the degree of the change that will occur. When realistic expectations are held, extension has been successful, rather than having been a failure.

Principle 27: Farmers need to feel valued

In NRM terms, Australia is asking its farmers to make a significant personal investment for what is largely a public benefit. Because of deeply-held notions of stewardship and the concept of good farm management, most Australian farmers are prepared to make their contribution. But they need to know that this contribution is appreciated and valued by the broader community. Although tax relief schemes tend not to benefit most farmers (as many have low taxable incomes) and many grant schemes don't necessarily achieve their intended objectives or have implementation and administration difficulties, some form of co-funding is important because it demonstrates to the farming community that the urban community cares and is prepared to pay for the environmental benefits they want. Evaluations of these schemes need to consider the effect of farmer commitment to NRM in general, and should not be evaluated strictly against narrow criteria.

Concluding thoughts

Farming is a social and cultural activity. Farm management practices are physical manifestations of cultural expression which are loaded with social meanings and significance, they are not solely technical. Farmers want practical advice, but that advice needs to be based on a social understanding of what farming is about. A key aspect of that social understanding is that diversity in agriculture should be conceived in social terms, rather than merely in physical or structural terms. Understanding farming from a social perspective will greatly assist in the promotion of sustainable agriculture.

While the language of the triple bottom line has had an influence in the world and there are merits to this approach, perhaps a better language to use, at least when it comes to farming, is embodied in the phrase coined by Andrew Campbell (Australia's first National Landcare Facilitator and sometime CEO of Land & Water Australia): the triple helix of landscapes, lifestyles and livelihoods (Vanclay 2004b). The words, landscapes, lifestyles and livelihoods, as well as the concept of a triple helix demonstrate the interwovenness, interactability and interdependence of these issues, a concept developed further in the chapter with Amabel Fulton. Agriculture is farming, and farming is a social activity. The process of enabling change in individuals, family farm businesses, communities and industries needs a greater understanding of the social nature of farming, extension and adoption processes.

Acknowledgements and dedication

This chapter is revised, expanded and updated from Vanclay F (2004) Social principles for agricultural extension to assist in the promotion of natural resource management. *Australian Journal of Experimental Agriculture* **44**(3), 213–222.

This chapter is dedicated to the Tasmanian Institute of Agricultural Research (TIAR) at the University of Tasmania. The AJEA paper on which this chapter is based was the first paper I wrote when I joined TIAR in 2002 and was based on the seminars I regarded as being the equivalent of my inaugural professorial address at the University of Tasmania. Furthermore, the revision of that paper for this book chapter (along with the co-editing of this book) is the last publication I prepared under TIAR's auspices prior to my move to the University of

Groningen in The Netherlands in mid-2010. While the messages in this chapter draw on my whole career rather than just my time in Tasmania, I believe they will continue to be valid for many years to come. Although I am mindful of the changing nature of extension agencies in Australia, I trust that these insights will be useful to TIAR and to other extension agencies throughout Australia and the world.

References

Carruthers G and Vanclay F (2007) Enhancing the social content of environmental management systems in Australian agriculture. *International Journal of Agricultural Resources, Governance and Ecology* **6**(3), 326–340.

Fleming A and Vanclay F (2010) Farmer responses to climate change and sustainable agriculture. *Agronomy for Sustainable Development* **30**(1), 11–19.

Geertz C (1963) *Agricultural Involution.* University of California Press: Berkeley.

Glyde S and Vanclay F (1996) Farming styles and technology transfer. In *Social Change in Rural Australia.* (Eds G Lawrence, K Lyons and S Momtaz) pp. 38–54. Central Queensland University: Rockhampton.

Howden P and Vanclay F (2000) Mythologisation of farming styles in Australian broadacre cropping. *Rural Sociology* **65**(2), 295–310.

Howden P, Vanclay F, Lemerle D and Kent J (1998) Working with the grain: farming styles amongst Australian broadacre croppers. *Rural Society* **8**(2), 109–125.

Kilpatrick S and Vanclay F (2005) Communities of practice for building social capital in rural Australia. In *Social Capital and Sustainable Community Development.* (Eds A Dale and J Onyx) pp. 141–158. UBC Press: Vancouver.

Lockie S, Mead A, Vanclay F and Butler B (1995) Factors encouraging the adoption of more sustainable crop rotations in south east Australia: profit, sustainability, risk, and stability. *Journal of Sustainable Agriculture* **6**(1), 61–79.

Mesiti L and Vanclay F (2006) Specifying the farming styles in viticulture. *Australian Journal of Experimental Agriculture* **46**(4), 585–593.

Phillips E and Gray I (1995) Farming practice as temporally and spatially situated intersections of biography, culture and social structure. *Australian Geographer* **26**(2), 127–32.

Quiggin J (1987) Land degradation: behavioural causes. In *Land Degradation: Problems and Policies.* (Eds A Chisholm and R Dumsday) pp. 203–212. Cambridge University Press: Sydney.

Rickson R, Saffigna P, Vanclay F and McTainsh G (1987) Social bases of farmers' responses to land degradation. In *Land Degradation: Problems and Policies.* (Eds A Chisholm and R Dumsday) pp. 187–200. Cambridge University Press: Sydney.

SELN (State Extension Leaders Network) (2006) 'Enabling change in rural and regional Australia: the role of extension in achieving sustainable and productive futures'. A discussion document produced by the State Extension Leaders Network August 2006. http://www.seln.org.au.

Silvasti T (2003) Bending borders of gendered labour division on farms: the case of Finland. *Sociologia Ruralis* **43**(2), 154–166.

Vanclay F (1986) Socio-economic correlates of adoption of soil conservation technology. Master's thesis. Department of Anthropology and Sociology, University of Queensland: St Lucia.

Vanclay F (1992a) The social context of farmers' adoption of environmentally sound farming practices. In *Agriculture, Environment and Society.* (Eds G Lawrence, F Vanclay and B Furze) pp. 94–121. Macmillan: Melbourne.

Vanclay F (1992b) Barriers to adoption: a general overview of the issues. *Rural Society* **2**(2), 10–12.

Vanclay F (1994) The sociology of the Australian agricultural environment. PhD thesis. Wageningen University: The Netherlands.

Vanclay F (1997) The social basis of environmental management in agriculture. In *Critical Landcare*. (Eds S Lockie and F Vanclay) pp. 9–27. Centre for Rural Social Research, Charles Sturt University: Wagga Wagga.

Vanclay F (2003) The impacts of deregulation and agricultural restructuring for rural Australia. *Australian Journal of Social Issues* **38**(1), 81–94.

Vanclay F (2004a) Social principles for agricultural extension to assist in the promotion of natural resource management. *Australian Journal of Experimental Agriculture* **44**(3), 213–222.

Vanclay F (2004b) The triple bottom line and impact assessment: how do TBL, EIA, SIA, SEA and EMS relate to each other? *Journal of Environmental Assessment Policy and Management* **6**(3), 265–288.

Vanclay F (2008) Place matters. In *Making Sense of Place*. (Eds F Vanclay, FM Higgins and A Blackshaw) pp. 2–11. National Museum of Australia Press: Canberra.

Vanclay F and Cary J (1989) *Farmer Perceptions of Dryland Soil Salinity*. School of Agriculture and Forestry, University of Melbourne: Melbourne.

Vanclay F and Lawrence G (1994) Farmer Rationality and the adoption of environmentally sound practices: a critique of the assumptions of traditional agricultural extension. *European Journal of Agricultural Education and Extension* **1**(1), 59–90.

Vanclay F and Lawrence G (1995) *The Environmental Imperative: Ecosocial Concerns for Australian Agriculture*. Central Queensland University Press: Rockhampton.

Vanclay F and Lockie S (1993) *Barriers to the Adoption of Sustainable Crop Rotations*. Centre for Rural Social Research, Charles Sturt University: Wagga Wagga.

Vanclay F, Leith P and Fleming A (2009) Understanding farming community concerns about adapting to a changed climate. In *Interdisciplinary Aspects of Climate Change*. (Eds W Filho and F Mannke) pp. 229–244. Peter Lang: Frankfurt.

Vanclay F, Mesiti L and Howden P (1998) Styles of farming and farming subcultures: appropriate concepts for Australian rural sociology? *Rural Society* **8**(2), 85–107.

Vanclay F, Howden P, Mesiti L and Glyde S (2006) The social and intellectual construction of farming styles: testing Dutch ideas in Australian agriculture. *Sociologia Ruralis* **46**(1), 61–82.

Vanclay F, Silvasti T and Howden P (2007) Styles, parables and scripts: diversity and conformity in Australian and Finnish agriculture. *Rural Society* **17**(1), 3–18.

5

Identifying potential adopters of an agricultural innovation

Geoff Kaine, Vic Wright, Ray Cooksey and Denise Bewsell

Summary

The return to public investment in agricultural research depends, in part, on the extent to which primary producers adopt the products of that research. Thus, maximising the return to investment in agricultural research involves identifying what research products are likely to be adopted by primary producers and by how many, and determining what processes are required to ensure the diffusion of research products among producers is as rapid as possible. In effect, adoption can be better targeted by identifying the relevant population of potential adopters and determining what processes will accelerate diffusion through that population. In this chapter, a method for identifying the population of potential adopters of an agricultural innovation is described. In essence, it is a process for discovering how agricultural innovations contribute to satisfying the needs of managers of agricultural enterprises. The method allows the population of potential adopters to be classified into benefit segments on the basis that producers with different farm contexts obtain different benefits from an innovation. The method is illustrated by application to a case study in horticulture. The fit of the method with the major paradigms in agricultural extension is discussed.

Introduction

We describe a theoretical framework for discovering how agricultural innovations contribute to satisfying the needs of primary producers as managers of agricultural enterprises. The framework draws on consumer behaviour theory and farming systems theory. It is based on the assumption that the adoption of agricultural innovations is a critically important decision for producers, and on the hypothesis that the benefits to be had from adopting an agricultural innovation are influenced by particular elements in farming systems that are specific to each innovation. The theoretical framework provided the basis for us to develop a method for properly specifying the population of potential adopters of agricultural innovations. The method allows the population of potential adopters to be classified into market segments on the basis that producers with different farm contexts obtain different benefits from an agricultural innovation.

Predicting the adoption of agricultural innovations

The population of potential adopters of an innovation can be defined as the set of decision makers who would perceive an innovation as offering a relative advantage, given sufficient knowledge of the consequences of adopting the innovation (Pannell *et al.* 2006; Rogers 1995). In other words, the population of potential adopters of an agricultural innovation consists of the set of producers who would perceive an innovation as offering a technical improvement of a type that is relevant to their needs and that is sufficiently compatible with their values, experiences and needs to merit adoption – given sufficient knowledge of their farm system and the consequences of adopting the innovation (Kaine 2008).

The population of potential adopters of an innovation represents the market for an innovation; that is, the potential buyers of the innovation. This means decisions to adopt new technologies and practices are purchase decisions. This is the case even when innovations are freely available because the acquisition and implementation of an innovation necessarily requires the investment of resources in making decisions about the uncertain benefits and costs of reconfiguring the farming system to accommodate the innovation (Byerlee *et al.* 1982; Collinson 2001; Crouch and Chamala 1981; Dillon and Heady 1958; Dorward *et al.* 2003).

Primary producers, however, are not specialist purchasers in the sense that organisational buyers are (Assael *et al.* 1990). Consequently, the literature on organisational buyer behaviour is largely irrelevant to the adoption of innovations in agriculture. The literature on decision making by non-specialist purchasers must suffice; this is the literature on purchasing by consumers (Assael *et al.* 1990; Kotler 2003; November 1984).

Consumer purchase behaviour and involvement

The literature on consumer purchasing is founded on social psychology which recognises that different decision processes are invoked in different circumstances (Derbaix and Vanden Abeele 1985; Krugman 1965; Levy 2005; Olson and Zanna 1993; Petty *et al.* 1983). This is important as the adoption of a novel agricultural technology or practice is qualitatively different from the routine purchase of familiar, unexceptional agricultural inputs (Derbaix and Vanden Abeele 1985). Central to explaining and predicting the different types of processes followed by consumers when making purchase decisions is the concept of involvement. Consumer involvement has been shown to influence purchase behaviours such as extensiveness of decision making, interest in advertising, brand commitment, frequency of product use, shopping enjoyment and social observations of product use and brand use (Mittal and Lee 1989).

Involvement is multidimensional and refers to the relative strength of the consumer's cognitive structure related to a focal object, a product (O'Cass 2000). In other words, consumer involvement is a motivational state that results from the consumer's perceptions that a product or activity can contribute to satisfying their goals, which may be utilitarian, social or hedonic (Mittal and Lee 1989). Involvement is intensified by perceived risk (Conchar *et al.* 2004; Dholakia 2001; O'Cass 2000). Perceived risk concerns the potential for undesirable outcomes and has three distinctive dimensions, psychological, social and functional (Dholakia 2001).

We propose that the decision to adopt an agricultural innovation is a high involvement purchase decision to the degree that the adoption of an agricultural innovation is perceived by primary producers as possibly contributing to satisfying their utilitarian, social, or hedonic goals as managers of agricultural enterprises (Fairweather and Keating 1994; Frost 2000; Gasson 1973; Ondersteijn *et al.* 2003). To the extent that the adoption of an agricultural innovation is also perceived by producers to carry risks of undesirable psychological, social and functional outcomes, their degree of involvement will be intensified.

High involvement purchases are characterised by extensive search for information about, and comprehensive consideration of, the attributes of the product and how these relate to the source of involvement. Assael (1998), Klonglan and Coward (1970), Parthasarathy *et al.* (1995), Rogers (1995) and others have proposed a variety of models that describe purchase decisions when involvement is high. The central theme in these models is that the process of making a purchase decision requires the development of criteria based on the consumer's motivation for involvement which are used to evaluate the alternative products or brands on offer (Assael 1998; Engel *et al.* 1995; Howard and Sheth 1969; Percy and Rossiter 1992).

The purchase criteria used by the consumer to evaluate products or brands represent the key benefits sought by the consumer and exemplify their usage or consumption situation (Assael 1998). For example, economy, dependability and safety may be key purchase criteria for many consumers with families that are buying motor vehicles to be used daily to transport family members, especially children. Having settled on a set of criteria for deciding between products, the consumer then evaluates the various makes and models against the criteria and makes a decision to purchase.

Consumers can be grouped into market segments on the basis of similarities and differences in the purchase criteria that they use to evaluate a product. Different segments will value products differently and/or on the basis of different product attributes. Knowledge of the criteria that will be used by consumers in a segment can be employed to tailor products to meet the specific needs of consumers in that segment and promote products accordingly (Kotler 2003).

Given that primary producers find the adoption of agricultural innovations highly involving, they can be expected to develop purchase criteria for evaluating innovations. Such criteria represent the key benefits producers are seeking in adopting an innovation. Further, the decision to adopt an innovation depends on their evaluation of the characteristics of the innovation based on their purchase criteria.

Producers' purchase criteria – the key benefits they seek from an innovation – should reflect their usage or consumption situation. The consumption situation for primary producers is defined by the characteristics of their farming system that influence the benefits to be had from adopting an agricultural innovation. We term these characteristics the farm context for the innovation.

The farm context for an innovation

Farms are systems made up of components such as resources (including people), exogenous environmental, social and economic constraints (such as access to markets), agricultural technology and management practices, and strategies for managing risks and pursuing objectives (Haines 1982; Norman 1980). These components and the relationships between them are used by the producer to produce agricultural outputs and realise family and business objectives (Norman 1980).

The presence of relationships between components means the components interact and they cannot be modified without causing a related change elsewhere in the system (Haines 1982). In other words, the components of a farming system are functionally related in that the state of one component potentially influences the states of other components (Ackoff 1971; Janssen and van Ittersum 2007). The relationships between components in a farm system may extend over space and time, may be direct or indirect, may be reciprocal, and may be nonlinear or stochastic.

The relationships between the components of a farm system restrict the way in which the components can be configured to achieve the objectives of the primary producer. In other words, the relationships between components place practical constraints on the ways in which a farm system can be managed within the restrictions created by the resources, technologies and practices that are available, and the strategies required to manage risks (Dorward *et al.* 2003; Gebremedhin and Swinton 2003). Consequently, the benefits to be had from introducing an innovation into a farm system will depend on precisely how the innovation changes the practical constraints on the way a farm system can be managed to achieve the objectives of the producer (Carruthers 2007; Cary *et al.* 2001; Collinson 2001; Cramb 2005; Crouch and Chamala 1981; Haines 1982).

These considerations suggest that a correct description of the consumption situation for an agricultural innovation requires the identification of those components and relationships within a farm system that are functionally related to the innovation. It is these components and relationships, which we call the 'farm context', that influence the benefits to be had from the innovation. These considerations also suggest that the components and relationships in the farm system that are functionally related to an innovation are the fundamental sources of the purchase criteria used by producers to evaluate the innovation. Hence, the farm context for an innovation denotes the components and relationships in farm systems that are causally related to the innovation and so shape the relative advantage the innovation offers.

This definition of farm context provides the basis for identifying the population of potential adopters of an agricultural innovation. This population was defined earlier as the set of decision makers who would perceive an innovation as offering the prospect of better achieving their goals given sufficient knowledge of the consequences of adopting the innovation. Given the definition of farm context, the population of potential adopters of an agricultural innovation can now be defined as the set of producers who would perceive the innovation as offering the prospect of better achieving their goals given sufficient knowledge of their farm context and, therefore, the consequences of adopting the innovation. The segments comprising potential adopters will be defined on the basis of different valuations of the innovation arising from different farm contexts.

The population of potential adopters is defined by the farm contexts for that innovation. It is independent of the knowledge and skills that may be needed to evaluate and implement the innovation correctly. This is because awareness of the innovation and possession of the knowledge and skills to evaluate and implement an innovation correctly are necessary but not sufficient conditions for adoption. Both conditions must be accompanied by the perception that the innovation offers relative advantage for adoption to occur. Awareness of an innovation and the acquisition of knowledge and skills that enable the innovation to be evaluated and implemented only influence the rate of adoption of the innovation.

An approach to identifying the population of potential adopters

In this section we propose a method for identifying and quantifying the population of potential adopters of an agricultural innovation. Broadly speaking, the method involves two stages. The objective in the first stage is to qualitatively identify the elements in farm systems that influence the benefits to be had from adopting a particular innovation and which thereby form the salient farm contexts for the innovation. The objective in the second stage is to quantify the proportion of producers in the population possessing the salient farm contexts for the innovation. This proportion is an estimate of the population of potential adopters of the innovation.

Qualitative stage

The method is based on three propositions that follow from the theoretical framework. The first is that the benefits to be had from adopting agricultural innovations depend on farm context. This means that knowledge of a producer's farm context is the basis for predicting whether a producer is a potential adopter of an innovation and why.

The second proposition is, given the adoption of innovations is a high involvement decision for producers, that their reasons for adopting innovations – their purchase criteria – will mirror their farm context. It follows from this proposition that producers with similar farm contexts will advance similar reasons for adopting an innovation. It also follows that producers from different farm contexts that adopt an innovation will offer different reasons for their behaviour.

The third proposition is that producers are the most authoritative source of knowledge about their farm contexts. This proposition means that, among all observers, producers have the richest understanding of their farm systems and the likely consequences of changing that system. Hence, producers will be the best source of information about the likely consequences of introducing an innovation into their farm systems and the factors that shape those consequences.

Together these three propositions lead to the conclusion that the set of farm contexts for an agricultural innovation can be discovered by using an interview process to elicit information from primary producers about their reasons for adopting the innovation. Given that similarities and differences in the reasoning supplied by producers ought to reflect similarities and differences in their farm contexts, then a dialectical approach to interviewing should be followed (Dick 1999). Hence, the first stage of the method.

Given that the elements in a farm system that constitute the farm context for an innovation are to be discovered through the elicitation process, the process cannot be structured *a priori* in terms of content or sample design. Consequently, a process is required that allows the elements that constitute farm context to emerge through disclosure, testing and confirmation. The process should also allow the sampling strategy to be refined as the elements that constitute farm context emerge.

These considerations suggest that, among the dialectical approaches to elicitation, convergent interviewing (Dick 1999) would be the most appropriate technique for identifying farm contexts. Convergent interviewing essentially involves a sequence of interviews with the information in each interview analysed and interpreted for consistency with previous interviews.

Quantitative stage

In the second stage of the method, the proportion of producers with farm systems that are consistent with the farm context for the innovation is quantified. This proportion is an estimate of the population of potential adopters of the innovation.

Through the interview process in the first stage, the various elements of the farm system that are thought to form the set of farm contexts for an innovation are identified. The benefits that the innovation is thought to generate in each farm context are also identified. Hence, the interview process yields a set of hypothesised associations between the various elements that constitute the set of farm contexts for an innovation, the adoption of the innovation and the benefits of the innovation. In principle then, these hypothesised associations may be tested statistically by gathering quantitative data on the elements that form the farm contexts and data on the adoption of the innovation. Such data could be gathered, for instance, by distributing a mail questionnaire to a random sample of producers based on the findings from the interviews subject to appropriate design and piloting.

The responses to such questionnaires could be used in a variety of ways. For example, producers could be classified using statistical techniques into segments representing the set of farm contexts for an innovation and the association between farm contexts for an innovation and the frequency of adoption of the innovation could be statistically tested. Also, having classified producers into a set of farm contexts, the relationship between farm context and the benefits to be had from an innovation could be tested. Associations between the adoption of the innovation, farm context and demographic characteristics or traits such as innovativeness could be tested to identify rival explanations for behaviour. Lastly, estimates could be made of the proportion of potential adopters in each segment and the potential population of adopters of the innovation as whole, given the sample of producers is statistically representative of the population of producers.

As a final point, a simple test of the soundness of the methods used to identify and quantify farm contexts may be conducted, in principle, by interviewing a sample of the producers who participated in the survey. These interviews would involve using laddering techniques to elicit producers' reasoning and decision making in regard to the innovation of interest and checking the concordance between their explanations of their decisions and the explanations predicted by their membership of a farm context segment.

The application of both stages of the method is illustrated in the following example in which different market segments are identified with respect to the adoption of micro-irrigation in the horticulture industry.

Predicting the adoption of micro-irrigation technologies in horticulture[1]

At the time this case study was conducted, best management practices were being identified for horticultural irrigation in northern Victoria and southern NSW. These practices included irrigation scheduling, nutrient management, salinity control and vigour management. Micro-irrigation was a prerequisite for many of these best management practices. Consequently, information identifying the factors that influenced the adoption of micro-irrigation was sought to assist in developing extension strategies promoting more widespread adoption of best management practices among fruit growers.

Qualitative stage

Convergent interviewing was used to identify the elements in the farming systems of fruit growing enterprises that defined the farm context for adopting micro-irrigation. In-depth, unstructured personal interviews were conducted with approximately 30 fruit growers. The growers who were interviewed were nominated by research and extension staff of the then Victorian Department of Natural Resources and Environment and NSW Agriculture. These growers were selected because they used different irrigation systems, they used different methods for scheduling irrigation, and they represented a variety of locations and enterprises of different scales.

The interviews revealed that irrigation management was a highly involved issue for fruit growers. The management of irrigation was a critical determinant of farm production and income, and could also influence their lifestyle. Growers must simultaneously manage a number of key activities including spraying, picking, grading and marketing in conjunction with irrigation. This placed a premium on their time. Typically, fruit blocks take between six and 10 hours to flood irrigate. Consequently, growers endeavour to concurrently irrigate as many fruit blocks as possible to reduce the amount of time spent irrigating. Blocks must also be given two to three days to dry out after irrigating before spraying or picking commences. As a result, on orchards with flood irrigation, operations such as spraying and picking are undertaken during the week and irrigation carried out on weekends, with all blocks on the farm being irrigated. This means that growers with flood irrigation have little opportunity to vary their irrigation routine, especially once picking has commenced.

Growers with impact sprinkler irrigation (under-tree knocker sprinklers) have more flexibility to vary their irrigation routine than growers with flood irrigation. Impact systems are pressurised, relatively high volume, controlled-flow systems. Because impact sprinklers are controlled-flow systems, there is no need to continuously monitor each block while irrigating and the water delivered to each block can be controlled by varying the irrigation period. Consequently, growers with impact sprinkler irrigation have the potential to irrigate blocks sequentially and to customise the irrigation of each block. This means irrigation on one block need not interfere with activities such as spraying and picking on another block. However, growers with impact sprinkler irrigation must water the entire block. Consequently, blocks must be left to dry out after being irrigated before activities such as spraying and picking can be undertaken.

The extent to which the potential benefits of impact sprinkler irrigation can be realised depends on the supply reliability of the irrigation distribution system. Growers serviced by spur channels, for example, may experience difficulties obtaining continuous supplies of moderate volumes of water over a sustained period. In these circumstances, growers with impact sprinkler irrigation may be forced to operate in a fashion similar to growers with flood irrigation, ordering large volumes and irrigating blocks concurrently to complete irrigation as rapidly as possible.

Growers with micro-irrigation indicated that they had the most flexibility to vary their irrigation scheduling. Micro-irrigation systems are pressurised, relatively low volume, controlled flow systems. As a result, there is no need to continuously monitor each block while irrigating and the water delivered to each block can be controlled by varying the irrigation period. This means that growers with micro-irrigation have the flexibility to irrigate blocks sequentially and to customise the irrigation of each block. Irrigation water is delivered directly to the base of the fruit tree by micro-irrigation systems. This means less water is required to irrigate a block of trees. It also means that, as blocks are not flooded, other activities such as spraying and picking can be undertaken concurrently with irrigation. Micro-irrigation is often the only form of irrigation suitable for orchards on hilly or sandy country.

As water is delivered directly to the base of the fruit tree, these micro-irrigation systems may be more effective than flood or impact systems on blocks using high density planting techniques such as the Tatura Trellis. These techniques involve hilling the soil under trees which prevents flood irrigation, and arranging trees in a trellis formation which prevents the use of impact sprinklers. Micro-irrigation systems are also more effective than flood or impact systems on blocks planted to dwarf rootstocks.

The interviews revealed growers were most likely to use micro-irrigation to reduce water use when they needed to manage a watertable or salinity problem, or were experiencing difficulties with the supply of irrigation water.

Quantitative stage

A mail survey based on the findings from the interviews was developed, piloted and distributed to fruit growers in northern Victoria and southern NSW. On the basis of the findings from the qualitative stage, the questionnaire was designed in three parts. The first part sought information on some basic characteristics such as orchard size, tree types, length of irrigation season and so on. The second part of the questionnaire was designed to elicit information on the irrigation systems used. These spanned aspects such as the type of irrigation system, area of orchard irrigated using each system, method of ordering irrigation water, and so on. Information was also sought on respondents' reasons for installing micro-irrigation systems. Respondents were asked to indicate which of the key factors that influence the adoption of micro-irrigation best described their reasons for installing micro-irrigation. In the third part of the survey, information was sought on the use of soil moisture monitoring systems.

Respondents were asked which of the key factors that influenced the adoption of monitoring best described their reasons for trying or using soil moisture monitoring.

Questionnaires were sent to all fruit growers in the Shepparton, Cobram and Swan Hill districts of Victoria and the Tumut and Batlow districts of NSW. The population of growers in these districts was approximately 780 650 in Victoria and 130 in NSW. The questionnaires were mailed in May 2000 with a reminder posted four weeks later. The case study and survey were also publicised through the local print media and industry newsletters. Within 10 weeks after the initial mailing, 251 questionnaires had been returned. This represented a response rate of 34% after allowing for incorrect addresses and for people who were no longer fruit growers.

An analysis of the sample revealed that the average area of orchards in each district was not significantly different across districts. However, the percentage of orchards in each district planted to different types of fruit trees was significantly different reflecting different growing environments. The percentage of orchard area under the different types of irrigation systems was also significantly different across districts. Flood irrigation in Shepparton and Swan Hill was based on bay or furrow irrigation whereas under-tree knocker sprinklers were the predominant form of irrigation in Cobram. The topography in Tumut and Batlow does not suit flood irrigation. Hence, micro-irrigation was the only form of irrigation in these two districts. The period over which growers in each district reported they were irrigating matched the experience of extension staff and provides a measure of confidence in the information supplied by the respondents.

Respondents who indicated they used a micro-irrigation system were asked to select from a list of statements those which best described their reasons for adopting micro-irrigation (Table 5.1). Five variables were constructed from the responses to these questions and used to classify fruit growers into market segments for micro-irrigation.[2]

For the purpose of the classification analysis, irrigating with under-tree knocker sprinklers was categorised as a type of flood irrigation, along with bay or furrow irrigation. We treated irrigating with under-tree knocker sprinklers as a form of flood irrigation system because, like bay and furrow irrigation, it imposes constraints on the conduct of other orchard activities such as spraying and picking because each of these systems involves wetting the entire orchard floor.

A scree test indicated that five market segments were present in the sample. Subsequent inspection of the segments confirmed the five segments were qualitatively different and that further subdivision was not substantively meaningful. The formation of the five segments is illustrated in Figure 5.1 and the profiles of the respondents in each segment, in terms of their reasons for adopting micro-irrigation, are presented in Table 5.2.

The benefit of installing micro-irrigation for respondents in the first segment (control and time-saving redevelopers) was to irrigate orchards that had been redeveloped to high density planting or plantings on trellis. These respondents also obtained benefits in regard to saving time and labour irrigating, and increasing flexibility in terms of picking and spraying fruit while irrigating. The growers in this segment represented approximately 23% of respondents. The benefits of converting orchards to micro-irrigation for respondents in the second segment (time-saving converters) were to save time and labour irrigating and to increase their flexibility in terms of picking and spraying fruit while irrigating. The growers in this segment represented approximately 24% of respondents. The respondents in the third segment (water-saving micro-irrigators) had installed micro-irrigation to manage limited water supplies or because micro-irrigation best suited their soils or topography. The respondents in this segment represented 17% of the sample and approximately half were from Batlow and Tumut. The respondents in the fourth segment (control redevelopers) had installed micro-irrigation to irrigate

Table 5.1.   Classification questions for irrigation systems in horticulture.

Questions
Have you installed micro-irrigation (i.e. trickle, drip, micro-jet or mini-sprinkler)? Which of the following reasons best describes why:

a)   I have been redeveloping blocks to closer plantings or plantings on trellis

b)   I have converted to micro-irrigation to save on time and labour spent irrigating

c)   I have been installing micro-irrigation because it increases my flexibility in terms of picking and spraying

d)   I installed micro-irrigation because of tree health problems due to groundwater, high watertables, or salinity

e)   I have installed micro-irrigation because of problems getting water delivered (e.g. because my orchard is on a spur channel)

f)   I installed micro-irrigation because it best suits the soil types/topography of my orchard

g)   I installed micro-irrigation because I have limited water supplies

h)   Other

Notes: The answers to the question on installing micro-irrigation are dichotomous (1=yes, 0=no). The responses to the reasons for installing micro-irrigation question were disaggregated to create seven separate dichotomous variables (1=yes, 0=no), one variable for each alternative.

(Source: Kaine and Bewsell 2002a, p. 9.)

orchards they had redeveloped to high-density plantings or plantings on trellis. The respondents in this segment represented approximately 15% of the sample. The growers in the fifth segment (flood irrigators) had not installed micro-irrigation and used bay or furrow flood irrigation or under-tree knocker sprinklers. These growers represented 22% of the sample.

The percentage of growers from the different irrigation segments in each district is reported in Table 5.3. In Shepparton, the majority of respondents that had installed micro-irrigation belonged to segments one or two. This suggested that the main motivations for installing

Table 5.2.   Market segment profiles for adoption of micro-irrigation.

Segments	1	2	3	4	5
Percentage of all respondents	23	24	17	15	22
Reasons for adopting*					
Redeveloped blocks or planted on trellis	100	0	25	76	0
Converted to save on time and labour spent irrigating	100	100	0	0	0
Installed to increase flexibility in orchard	92	65	12	21	0
Installed because of tree health problems	32	31	8	0	0
Installed because of problems getting water delivered	4	9	0	0	0
Installed because it best suited soil or topography	36	25	62	0	0
Installed because of limited water supplies	25	20	54	0	0

Notes:

* Numbers are percentage of respondents in each segment nominating reason.

Segments: 1 – control and time-saving redevelopers; 2 – time-saving converters; 3 – water-saving micro-irrigators; 4 – control redevelopers; 5 – flood irrigators.

(Source: Adapted from Kaine and Bewsell 2002a, pp. 12–13.)

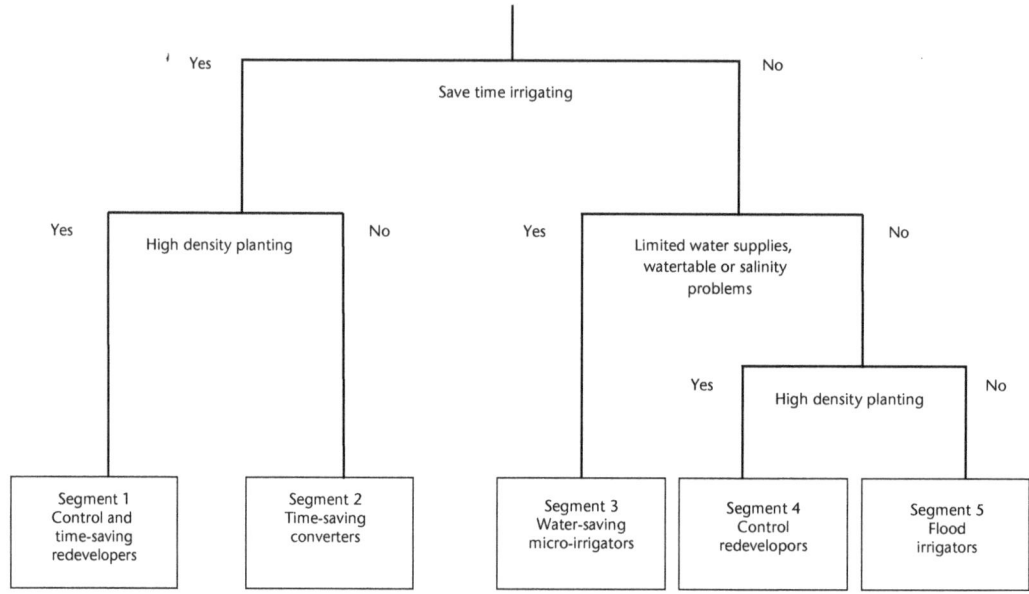

Figure 5.1. Classification of market segments for irrigation in horticulture.

(Source: Kaine and Bewsell 2002a, p. 11.)

micro-irrigation in the Shepparton district were to save time irrigating, to increase flexibility in managing orchard activities and to redevelop orchards.

In Cobram, a relatively high percentage of respondents used under-tree knocker sprinklers to flood irrigate. Those respondents who had installed micro-irrigation were classified into segment four. This indicated the installation of micro-irrigation in the Cobram district was mainly the result of redeveloping orchards to high-density plantings or plantings on trellis.

A relatively high percentage of respondents in Swan Hill used furrow irrigation. Those respondents in Swan Hill who had installed micro-irrigation were classified into the first and second market segments. This suggests that respondents in Swan Hill, like those in Shepparton, had installed micro-irrigation to save time and labour, to increase flexibility in managing orchard activities and when redeveloping orchards.

Table 5.3. Characteristics of market segments for micro-irrigation.

Segments	1	2	3	4	5
Shepparton district[1]*	30	28	11	12	19
Cobram district[2]*	10	10	17	34	29
Swan Hill district[3]*	26	27	2	9	36
Tumut and Batlow*	14	20	57	9	0

Notes:

Numbers are percentage of respondents in each region except for use of irrigation consultants and services where numbers are percentage of respondents in each segment.

* denotes a statistically significant difference at 0.05 using Pearson's chi-square test.

[1] includes East Shepparton, Ardmona, Tatura and Kyabram.

[2] includes Invergordon.

[3] includes Tresco, Woorinen and Nyah.

Segments: 1 – control and time-saving redevelopers; 2 – time-saving converters; 3 – water-saving micro-irrigators; 4 – control redevelopers; 5 – flood irrigators.

(Source: Adapted from Kaine and Bewsell 2002a, pp. 15–16.)

In Tumut and Batlow most respondents had installed micro-irrigation to save water. Consequently, respondents in this district were classified into segment three.

Extension

Extension strategies and priorities were formulated in a workshop with Department staff involved in horticultural extension. In the workshop, the results of the case study were combined with the knowledge and experience of Department staff using a program logic approach (Mayeske 1994). The results indicated that respondents in the different market segments for micro-irrigation had different information needs, but there were some unifying themes from an extension perspective.

Workshop participants identified a number of problems that might cause fruit growers with flood or micro-irrigation to be dissatisfied with their irrigation management. The response to most problems with flood irrigation, such as a shortage of labour or high watertables, was to install micro-irrigation. Consequently, the problems that were identified in the workshop concerned difficulties either in converting from flood to micro-irrigation or difficulties in managing micro-irrigation systems once installed. In response, extension strategies were formulated in relation to assisting growers to:

- manage changing from flood irrigation to micro-irrigation of young trees if they were redeveloping their orchards
- manage changing from flood irrigation if they were converting established trees to micro-irrigation
- make decisions in relation to the choice and design of micro-irrigation systems
- manage and maintain micro-irrigation systems and deal with problems such as high watertables or excessive tree vigour.

The extension strategies for each irrigation market segment are summarised in Table 5.4. The extension strategies were validated by conducting a final round of unstructured interviews with fruit growers from the different market segments in the Shepparton district. The growers were identified from the contact details they supplied when completing the survey questionnaire. These interviews were also used to verify the classification of the growers into market segments for micro-irrigation.

The validation interviews confirmed and reinforced the earlier findings of the case study and the context-specific priorities for extension. All growers who were interviewed indicated that they had experienced some initial problems with micro-irrigation. The growers also indicated they had overcome these problems within a season or two through learning by trial and error and through discussions with other growers and irrigation equipment suppliers. None

Table 5.4.   Extension strategies developed for irrigation system market segments.

Segments	1	2	3	4	5
Managing young trees[2]	×			×	×
Managing established trees[2]		×	×		×
Working with designers to ensure good system design[2]	×	×	×	×	×
Technical advice[1,2]	×	×	×	×	×

Notes:
[1] Technical advice covers issues such as excessive tree vigour, hares chewing lines etc., particularly emphasising the links between fruit quality, vigour and irrigation management.
[2] Fruit growers in segment five should be targeted when converting to one of the other segments.
Segments: 1 – control and time-saving redevelopers; 2 – time-saving converters; 3 – water-saving micro-irrigators; 4 – control redevelopers; 5 – flood irrigators.

(Source: Kaine and Bewsell 2002b, p. 17.)

indicated that they had experienced any long-term problems and all were satisfied with their current irrigation system and management. These results were consistent with the views expressed by the growers that were interviewed in the first stage of the case study.

Conclusions

At the time this case study was conducted, best management practices were being identified to improve the efficiency of water use. Efficient water use, however, was not identified as a key benefit influencing the adoption of micro-irrigation by the majority of growers. Hence, extension efforts to promote the benefits of adopting micro-irrigation and associated best management practices should not have simply focused on highlighting gains in water use efficiency. The only region where this was particularly relevant was in Tumut and Batlow; however, most growers there had already adopted micro-irrigation to increase efficiency and save water.

For most growers, the key benefits influencing the adoption of micro-irrigation at the time were a need to reduce the time spent irrigating, a desire to increase flexibility in managing irrigation, spraying and harvesting activities, and a desire to increase productivity and profitability by redeveloping orchards to closer planting or trellis designs. This meant that the population of potential adopters of micro-irrigation was determined by factors that influenced these benefits. Such factors may have included changes in family composition which influenced labour availability, changes in the scarcity of irrigation water and the reliability of delivery of irrigation water, and the development of orchards in areas unsuited to furrow irrigation. Consequently, the role of extension in increasing the rate of adoption of micro-irrigation was to facilitate the process of changing from flood to micro-irrigation once circumstances prompted growers to make the change.

Discussion

The idea that the population of potential adopters of innovations can be defined by identifying the farm context for an innovation raises many theoretical and practical implications for agricultural research, extension and policy. We consider several of these, starting with some implications in regard to agricultural research.

Agricultural research

The first implication is that the population of potential adopters tends to be overestimated even where relatively rigorous procedures are used to identify such populations, as is the case with the identification of recommendation domains in farming systems research, for example. Our experiences suggest that the population of potential adopters of an innovation is often only a small proportion of the producers in a particular industry and region. Relatedly, the membership of the potential population of adopters and the segments within that population will change in response to changes in farm context. Changes in farm context may be the result of changes within the farm enterprise. For example, the retirement of a senior family member from active participation in orchard activities may trigger the need to adopt labour-saving technologies such as micro-irrigation technology. It would be easy, in this instance, to misinterpret the timing of adoption as evidence that the senior family member was conservative and preventing younger, apparently more innovative members of the family, from adopting the technology. Changes in farm context may also be the result of changes outside the farm enterprise; for instance, changes to regional irrigation infrastructure that may change the farm context of growers with respect to the supply of irrigation water and thereby trigger the adoption of soil moisture monitoring (Boland *et al.* 2005). The population of potential adopters of innovations will also change should the utilitarian, social and hedonistic objectives of producers change. For instance, changes in the lifestyle aspirations of producers may result in a desire to reduce the time and effort expended on farm activities, such as irrigation, leading to the adoption of labour-saving technologies such as automatic irrigation (Kaine and Bewsell 2000, 2004).

The second implication concerns the tailoring of research to target producers with different farm contexts. Producers from different market segments can use different choice criteria to evaluate an innovation and so may favour different attributes in an innovation. The method described here may help tailor the products of research to satisfy the different choice criteria that will be used by producers from different market segments better. Knowledge of the different farm contexts may assist farming systems researchers to define more precisely the point at which prototypes of innovations should be released to producers for subsequent modification and adaptation (Collinson 2001; Douthwaite *et al.* 2002; Sumberg 2005).

The third implication concerns the recruitment of producers as collaborators in research and development activities. The participation of producers is critical to the successful conceptualisation, development and commercialisation of innovations. The theory described here, however, supported by the case study, highlighted that the benefits producers seek from an innovation differ depending on their farm contexts. To the extent that producers from different contexts may favour different attributes in the innovation, researchers must be careful to ensure producers from each of the target farm contexts are recruited to collaborate in the development of the innovation. Hence, using the methods described here for identifying the set of farm contexts for an innovation may have merit as a means of ensuring collaborating producers are recruited from the appropriate farm contexts.

Agricultural extension

The first implication of the findings is that the rate of diffusion of agricultural innovations persistently tends to be underestimated because the population of potential adopters persistently tends to be overestimated. Typically, apparently low rates of adoption are equated with low diffusion. The findings that the population of potential adopters of an innovation is typically a proportion of the producers in a particular industry and region implies directly that apparently low rates of adoption are, in fact, higher when calculated with reference to the actual population of potential adopters. A relatively small number of adopters may actually represent a high level of diffusion if the population of potential adopters is relatively small. On the other hand, a relatively high number of adopters may actually represent a low level of diffusion if the population of potential adopters is relatively large. Similar logic can be applied to attempting to judge the success of extension programs by levels of attendance at events such as field days and workshops. The method provides a means of resolving these difficulties by providing a sound basis for judging relative rates of adoption.

The second implication concerns the tailoring of extension messages to target producers with different farm contexts. Given knowledge of market segments, extension messages may be tailored to appeal to the specific choice criteria used by producers in a market segment to evaluate an innovation. This offers the possibility that targeting extension activities may become a practical reality by enabling the alignment of extension messages and tailored research products with the choice criteria of producers in a segment and their associated preferences regarding the attribute of the innovation, respectively. Whether there is merit in doing so will depend on the relative size of segments, the value to each segment of the customised features, and the costs of tailoring the extension effort.

The final implication for extension of the findings that will be considered concerns group extension. The proposition that the population of potential adopters of an agricultural innovation is composed of market segments based on differences in farm contexts requires that managers of extension programs be sensitive to the ways in which these differences may influence the interactions between producers in group situations. For example, workshops involving producers from a variety of segments may be useful in producing lists of elements in the farm system that influence the benefits to be had from an innovation. However, such workshops must be managed carefully if the causal interactions between these elements across different farm contexts are to be described with sufficient clarity to reveal the contexts. Kaine *et al.* (2003) provide an example where workshops were used to identify the various elements in the farm context that influence producers' preferences for different mechanisms for administering animal health products. Producers identified a number of contradictory items. Subsequent application of the method for identifying the population of potential adopters resolved these contradictions by revealing that these elements functioned in different ways depending on the presence of other elements in the farm context.

Managers of extension programs must be sensitive to the ways in which differences in farm contexts might influence the interactions between producers in group situations in that there is no compelling reason to suppose that producers from different farm contexts will necessarily understand the reasons for their different perspectives. Misunderstandings among producers about the functioning of different farm contexts may partly explain producers' constructions of apparently fictional farming styles and their judgments about those styles (Howden and Vanclay 2000).

Agricultural policy

The first implication for agricultural policy concerns the limits to the effectiveness of extension as a policy instrument for promoting voluntary behaviour change. Extension can facilitate or accelerate voluntary change by reducing the effort and time producers need to spend in searching for information and acquiring skills. In other words, extension can assist agricultural innovations to disseminate through a population of potential adopters more rapidly than would otherwise be the case. As discussed above, however, extension cannot itself change the population of potential adopters except in unusual and problematic circumstances. This means that other policy instruments must be employed in conjunction with extension in circumstances where voluntary change by the population of potential adopters will not achieve a policy outcome.

The second and related implication is that policy makers should be sensitive to the distinction between policy instruments that accelerate the diffusion of agricultural innovations and policy instruments that change the population of potential adopters. The population of potential adopters is defined by the set of farm contexts for an agricultural innovation. Hence, the population of potential adopters of an agricultural innovation can be changed by appropriately changing the farm contexts for the innovation. Policy instruments that can change farm context and thereby the population of potential adopters include research to change the features of innovations, changes in property rights, changes in regulations and changes in public infrastructure and utilities. Each of these has the capacity to change the population of potential adopters by modifying the farm context of producers and thereby the benefits they are seeking and the choice criteria they use to evaluate an innovation. Examples include restrictions on pesticides (Jeger 2000; Kogan 1998), changes to regional irrigation infrastructure (Boland *et al.* 2005) and the regulation of effluent management (Davies *et al.* 2007).

Policy instruments that can change the rate of diffusion of agricultural innovations are extension, research, and incentives and charges. Research may change the rate of diffusion by offering innovations with features that are better tailored to the requirements of producers, or innovations that are more easily adapted to different farm contexts. Incentives and charges change the rate of diffusion by increasing or decreasing, respectively, the relative advantage offered by an innovation, given that the innovation offers a relative advantage independently of the incentive or charge.

The third implication is that the findings point to the potential for producers to respond in unexpected ways to policy initiatives. As discussed earlier, agricultural innovations may be adopted by producers to realise benefits other than those for which the innovations were designed. The characteristics of producers and farming systems are such that variety will emerge in the uses to which agricultural innovations may be put (Kaine and Higson 2006). It is possible that policies may be implemented to accelerate the diffusion of an innovation on the grounds that adoption will generate public benefits, but the innovation may be implemented in ways that are not consistent with achieving those public benefits (Kaine and Johnson 2004). The consequence is that producers may adopt and implement an innovation in ways that do not contribute as strongly as expected to the outcome sought by the policy maker. The result is a gap between the desired policy outcome and the outcome that is actually achieved. For example, incentives have been offered for construction of water storages on farms as a means of ensuring irrigation water is contained in a closed system on farms, thereby increasing irrigation efficiency and reducing accessions to watertables. The use of storages for this purpose requires that the storages be empty at the start of irrigation; however, producers in some farm contexts have constructed the storages in order to improve control over the timing of irrigation. This use requires the storages be full at the commencement of irrigation, thereby negating the purpose of the incentives (Kaine and Bewsell 2000, 2004).

Finally, the adoption of agricultural innovations by producers cannot be treated as a criterion for making the judgment that some producers are necessarily better managers of their farm business than others. The failure of producers to adopt an innovation most likely merely means that the innovation is unsuited to their farm context at that point in time. To judge such producers as incompetent or conservative without a proper understanding of their farm system is mistaken, misleading and unjustly pejorative. To claim that producers who have adopted innovations are superior to or more innovative than those who have not is, without proper understanding of their farm system, equally mistaken and misleading and, perhaps, a self-serving exaggeration.

These implications suggest that policy makers require a comprehensive understanding of the farm contexts for agricultural innovations if they are to accurately anticipate and influence the behaviour of producers in regard to the adoption of agricultural innovations.

Conclusion

We have described a theoretical framework for discovering how agricultural innovations contribute to satisfying the needs of primary producers as managers of agricultural enterprises. The framework provided a basis for us to develop a method for quantifying the population of potential adopters of agricultural innovations. This allowed the population of potential adopters to be classified into market segments on the basis that producers with different farm contexts obtain different benefits from an agricultural innovation. The method was illustrated using a case study in which market segments were identified for the adoption of micro-irrigation systems in horticulture. The results indicated that, at that time, the adoption of micro-irrigation by fruit growers in northern Victoria was motivated primarily by labour concerns. The results demonstrated that market segments can be identified based on differences in the benefits farmers seek from an innovation.

A number of critical implications follow. First, the decisions of producers to adopt an innovation (or not) are rational in that they are based on their private perceptions of the relative advantage offered (or not) by an innovation. Second, producers' perceptions of relative advantage need not necessarily involve consideration of the potential public benefits of adoption. Third, in the absence of a proper understanding of the differences in the farm systems of producers, value judgments about adopters and non-adopters are likely mistaken and misleading.

Endnotes

1   This section draws on Boland *et al.* (2005), Kaine *et al.* (2005) and Kaine and Bewsell (1999, 2002a, 2002b). This case study was financially supported by the Victorian Department of Natural Resources and Environment. Subsequent to this study an extended drought occurred, but it did not affect this research.
2   The classification analysis was conducted using a monothetic divisive clustering algorithm available in CLUSTAN (Wishart 1987) which is designed for use with dichotomous data. The similarity coefficient used was squared Euclidean distance.

References

Ackoff RL (1971) Towards a system of system concepts. *Management Science* **17**(11), 661–671.
Assael H (1998) *Consumer Behaviour and Marketing Action.* South Western College Publishing: Cincinnati.
Assael H, Reed PW and Patton M (1990) *Marketing: Principles and Strategy.* Harcourt Brace: Sydney.
Boland A., Bewsell D and Kaine G (2005) Adoption of sustainable irrigation management practices by stone and pome fruit growers in the Goulburn/Murray Valleys, Australia. *Irrigation Science* **24**(2), 137–145.
Byerlee D, Harrington L and Winkelmann DL (1982) Farming systems research: issues in research strategy and technology design. *American Journal of Agricultural Economics* **64**(5), 897–904.
Carruthers G (2007) Using the EMS process as an integrative farm management tool. *Australian Journal of Experimental Agriculture* **47**(3), 312–324.
Cary J, Webb T and Barr N (2001) *The Adoption of Sustainable Practices: Some New Insights. An Analysis of Drivers and Constraints for the Adoption of Sustainable Practices Derived from Research.* Land and Water Australia: Canberra.
Collinson M (2001) Institution and professional obstacles to a more effective research process for smallholder agriculture. *Agricultural Systems* **69**(1), 27–36.
Conchar MP, Zinkhan GM, Peters C and Olavarrieta S (2004) An integrated framework for the conceptualization of consumers' perceived-risk processing. *Journal of the Academy of Marketing Science* **32**(4), 418–436.
Cramb RA (2005) Farmers' strategies for managing acid upland soils in Southeast Asia: an evolutionary perspective. *Agriculture, Ecosystems and Environment* **106**(1), 69–87.
Crouch BR and Chamala S (Eds) (1981) *Innovation and Farm Development: A Multi-dimensional Model in Extension, Education and Rural Development.* Wiley and Sons: Brisbane.
Davies A, Kaine G and Lourey R (2007) 'Understanding factors leading to non-compliance with effluent regulations by dairy farmers'. Technical Report 2007/37. Environment Waikato: Hamilton, New Zealand.

Derbaix C and Vanden Abeele P (1985) Consumer inferences and consumer preferences. The status of cognition and consciousness in consumer behaviour theory. *International Journal of Research in Marketing* **2**(3), 157–174.

Dholakia UM (2001) A motivational process model of product involvement and consumer risk perception. *European Journal of Marketing* **35**(11/12), 1340–1360.

Dick B (1999) *Rigour without Numbers: The Potential of Dialectical Processes as Qualitative Research Tools*. Interchange: Chapel Hill.

Dillon JL and Heady EO (1958) Decision criteria for innovation. *Australian Journal of Agricultural Economics* **2**(2), 113–120.

Dorward P, Galpin M and Shepherd D (2003) Participatory farm management methods for assessing the suitability of potential innovations. A case study on green manuring options for tomato producers in Ghana. *Agricultural Systems* **75**(1), 97–117.

Douthwaite B, Keating JDH and Park JR (2002) Learning selection: an evolutionary model for understanding, implementing and evaluating participatory technology development. *Agricultural Systems* **72**(2), 109–131.

Engel JF, Blackwell RD and Miniard PW (1995) *Consumer Behavior*. Dryden: Florida.

Fairweather JR and Keating NC (1994) Goals and management styles of New Zealand farmers. *Agricultural Systems* **44**(2), 181–200.

Frost FM (2000) Value orientations: impact and implications in the extension of complex farming systems. *Australian Journal of Experimental Agriculture* **40**(4), 511–517.

Gasson R (1973) Goals and values of farmers. *Journal of Agricultural Economics* **24**(3), 521–537.

Gebremedhin G and Swinton SM (2003) Investment in soil conservation in northern Ethiopia: the role of land tenure security and public programs. *Agricultural Economics* **29**(1), 69–84.

Haines MR (1982) *Introduction to Farming Systems*. Longman House: Essex.

Howard J and Sheth J (1969) *The Theory of Buyer Behaviour*. Wiley: New York.

Howden P and Vanclay F (2000) Mythologization of farming styles in Australian broadacre cropping. *Rural Sociology* **65**(2), 295–310.

Janssen S and van Ittersum MK (2007) Assessing farm innovations and responses to policies: a review of bio-economic farm models. *Agricultural Systems* **94**(3), 622–636.

Jeger MJ (2000) Bottlenecks in IPM. *Crop Protection* **19**(8–10), 787–792.

Kaine G (2008) The adoption of agricultural innovations. Unpublished PhD thesis. University of New England: Armidale.

Kaine G and Bewsell D (1999) 'Soil monitoring, irrigation scheduling and fruit production (Part 1)'. School of Marketing and Management, University of New England: Armidale.

Kaine G and Bewsell D (2000) 'Irrigation systems, irrigation management and dairy farming (first report)'. School of Marketing and Management, University of New England: Armidale.

Kaine G and Bewsell D (2002a). 'Soil monitoring, irrigation scheduling and fruit production (Part 2)'. School of Marketing and Management, University of New England: Armidale.

Kaine G and Bewsell D (2002b) 'Soil monitoring, irrigation scheduling and fruit production (Part 3)'. School of Marketing and Management, University of New England: Armidale.

Kaine G and Bewsell D (2004) 'Irrigation systems, irrigation management and dairy farming (second report)'. School of Marketing and Management, University of New England: Armidale.

Kaine G and Higson M (2006) Understanding variety in landholders' responses to resource policy. *Australasian Agribusiness Review* **14**, Paper 6.

Kaine G and Johnson F (2004) 'Applying marketing principles to policy design and implementation'. Social Research Working Paper. AgResearch: Hamilton, New Zealand.

Kaine G, Tarbotton IS and Bewsell D (2003) 'Factors influencing choice of health products for livestock'. AgResearch: Hamilton, New Zealand.

Kaine G, Bewsell D, Boland A and Linehan C (2005) Using market research to understand the adoption of irrigation management strategies in the stone and pome fruit industry. *Australian Journal of Experimental Agriculture* **45**(9), 1181–1187.

Klonglan GE and Coward EWJ (1970) The concept of symbolic adoption: a suggested interpretation. *Rural Sociology* **35**(1), 77–83.

Kogan M (1998) Integrated pest management: historical perspectives and contemporary developments. *Annual Review of Entomology* **43**, 243–270.

Kotler P (2003) *Marketing Management.* Prentice Hall: New Jersey.

Krugman HE (1965) The impact of television advertising: learning without involvement. *The Public Opinion Quarterly* **29**(3), 349–356.

Levy SJ (2005) The evolution of qualitative research in consumer behaviour. *Journal of Business Research* **58**(3), 341–347.

Mayeske GW (1994) *Life Cycle Program Management & Evaluation: An Heuristic Approach.* United States Department of Agriculture Extension Service: Washington.

Mittal B and Lee M (1989) A causal model of consumer involvement. *Journal of Economic Psychology* **10**(3), 363–389.

Norman DW (1980) The farming systems approach: relevancy for the small farmer. *MSU Rural Development Papers* **5**, 10. Department of Agricultural Economics, Michigan State University: East Lansing.

November P (1984) *Practical Marketing in Australia.* John Wiley: Brisbane.

O'Cass A (2000) An assessment of consumers product, purchase decision, advertising and consumption involvement in fashion clothing. *Journal of Economic Psychology* **21**(5), 545–576.

Olson JM and Zanna MP (1993) Attitudes and attitude change. *Annual Review of Psychology* **44**, 117–154.

Ondersteijn CJM, Giesen GWJ and Huirne RBM (2003) Identification of farmer characteristics and farm strategies explaining changes in environmental management and environmental and economic performance of dairy farms. *Agricultural Systems* **78**, 31–55.

Pannell DJ, Marshall GR, Barr N, Curtis A, Vanclay F and Wilkinson R (2006) Understanding and promoting adoption of conservation practices by rural landholders. *Australian Journal of Experimental Agriculture* **46**(11), 1407–1424.

Parthasarathy M, Rittenburg TL and Ball AD (1995) A re-evaluation of the product innovation-decision process: the implications for product management. *Journal of Product and Brand Management* **4**(4), 35–47.

Percy L and Rossiter JR (1992) A model of brand awareness and brand attitude advertising strategies. *Psychology and Marketing* **9**(4), 263–274.

Petty RE, Cacioppo JT and Schumann D (1983) Central and peripheral routes to advertising effectiveness: the moderating role of involvement. *The Journal of Consumer Research* **10**(2), 135–146.

Rogers EM (1995) *Diffusion of Innovations.* 4th edn. Free Press: New York.

Sumberg J (2005) Constraints to the adoption of agricultural innovations. Is it time for a re-think? *Outlook on Agriculture* **34**(1), 7–10.

Wishart D (1987) *CLUSTAN User Manual.* University of St Andrews: Fife.

6

Identifying and targeting adoption drivers

Rick S Llewellyn

Summary

Information and learning are often the primary tools available for influencing practice change. In this chapter, the role of information and learning in landholder adoption decisions is explored with a view to providing a pathway for more effective extension planning and delivery. Where potential adopters are found to hold perceptions that differ from those held by comparable users of a practice, there is the potential for learning. Where those perceptions are also associated with a greater likelihood of adoption, extension that promotes that learning is likely to lead to more rapid and better-informed adoption decisions. Steps and examples are presented that show an approach to identifying perceptions that are influential in the adoption decision and have the potential to be targeted effectively.

Introduction

Many projects aim for extensive adoption and practice change, often within a very short period of time. Large amounts of public and private funding are spent on these projects which are usually only achieved by influencing a large number of independent adoption decisions among a highly diverse population of individual decision makers. Often, particularly in larger projects, not much is known about the individual decision makers who will determine whether the practice change targets are going to be met. Further, most projects only ever have personal contact with a small proportion of the potential adopters who will ultimately determine whether the project succeeds or fails.

Adding to the challenge, the tools available for a typical project to influence practice change by rural landholders are usually limited to those involving information and learning. The carrots and sticks of incentives and legislation are rarely at hand. Given the many diverse sources of information for farmers, a single project is only ever destined to be a bit-player in delivering information and promoting learning of the relative merits of adopting a particular practice (see Barr, Chapter 9 of this volume).

This chapter looks at ways to understand adoption decisions for particular innovations so that the most likely opportunities to assist decision making and influence adoption decisions can be identified and targeted. To do this requires a level of confidence so that at least some of the main factors influencing a substantial proportion of adoption decisions, or future adoption decisions, among the target population can be identified and understood, especially those having the potential to be influenced. Beginning with a brief background on the theoretical role of information and learning in adoption decisions, this chapter then looks at the steps for identifying key factors in adoption decisions, including examples where opportunities for targeted extension, information and learning have been found and tested.

Information and learning as adoption drivers

The potential for extension to influence learning changes as the diffusion process progresses. In the early phases of diffusion, when relatively little is known about a new practice, information and learning-related factors are major determinants of the speed of adoption (Feder and Umali 1993; Lindner 1987; Marsh *et al.* 2000). This is when there is a greater scope for extension to influence the rate of adoption.

Essentially, information can have the greatest impact on learning when the decision maker is not yet well-informed about the innovation and has a high level of uncertainty about the innovation's likely performance. The information gained contributes to a learning process through which decision makers adjust their perceptions. As examples, this can involve shifts in the perceived outcome or cost if a particular practice is used; reduced (or increased) uncertainty that a particular outcome will occur when used on-farm; or awareness of a new effect of the innovation that was not previously considered that may influence its overall relative advantage if adopted.

Innovations that offer greater relative advantage to decision makers are more likely to generate more positive messages about the value of adopting (Lindner 1987). When these advantages are highly observable, the likelihood that more will adopt earlier increases (Pannell *et al.* 2006). As diffusion proceeds, there is a greater likelihood that highly relevant information will be obtained at less or no cost as a result of observation and contact with multiple neighbours, for example.

An abundance of readily accessible local information is also likely to mean an abundance of information with perceived 'quality' or effectiveness in terms of being able to influence learning (Fischer *et al.* 1996; Marra *et al.* 2001). Local sources of information are likely to have greater effectiveness than more distant sources (Lindner *et al.* 1982). Being local means that uncertainty (and variance) surrounding the relevance and applicability of the information to the decision maker's own on-farm use is often reduced. Therefore, information produced by local use of a new innovation, or especially an on-farm trial, is likely to have higher 'quality' or value (Abadi Ghadim 2005; Llewellyn 2007). Similarly, perceived information quality and effectiveness can also be influenced by factors such as perceived validity of the source (Leathers and Smale 1991). As more information becomes more readily available there is greater opportunity for non-adopters' perceptions of particular features or components of the innovation to become more consistent with those held by adopters.

Considering all of these factors, it is argued that more can be done to target extension resources (and research and development) to improve its effectiveness. By adequately considering the information and learning aspects of the adoption decision, information and learning can be directed towards factors that (a) are influential in the adoption decision, and (b) can be influenced by extension. Decisions to adopt or not adopt can therefore be improved and accelerated. Improved targeting of extension has the potential to improve the cost-effectiveness not only of investments in extension but also of farmer time and resources spent participating in extension and learning. In the following section some suggested steps for doing this are presented together with examples.

Two examples of identifying adoption drivers

Explaining no-till adoption across Australian cropping regions

From a traditional cropping system involving multiple cultivations of typically fragile soil, the shift to no-tillage farming represents one of the most substantial landscape changes in modern Australian agriculture. Reduced tillage to decrease the private and public costs of soil erosion has been the subject of substantial extension investment over a long period and still remains an extension priority (e.g. Australian Government 2008). It also provides an interesting case study on opportunities to target adoption drivers.

The adoption of no-till across selected Australian grain producing regions is now well advanced and nearing peak adoption in many districts (see Figure 6.1). The results are based on a study of no-till adoption across selected cropping regions of Western Australia, South Australia, Victoria, New South Wales and southern Queensland involving interviews with 1170 grain growers in 2008 (Llewellyn and D'Emden 2010). The proportion of growers in the sample using at least some no-till is now peaking at levels nearing 90% in some regions. Based on the year of first use, as stated by grain growers in 2008, the curves display the form of a classical diffusion curve; a long lead time with slow or no adoption followed by a steady increase before a rapid surge in adoption, before slowing as likely peak adoption is approached.

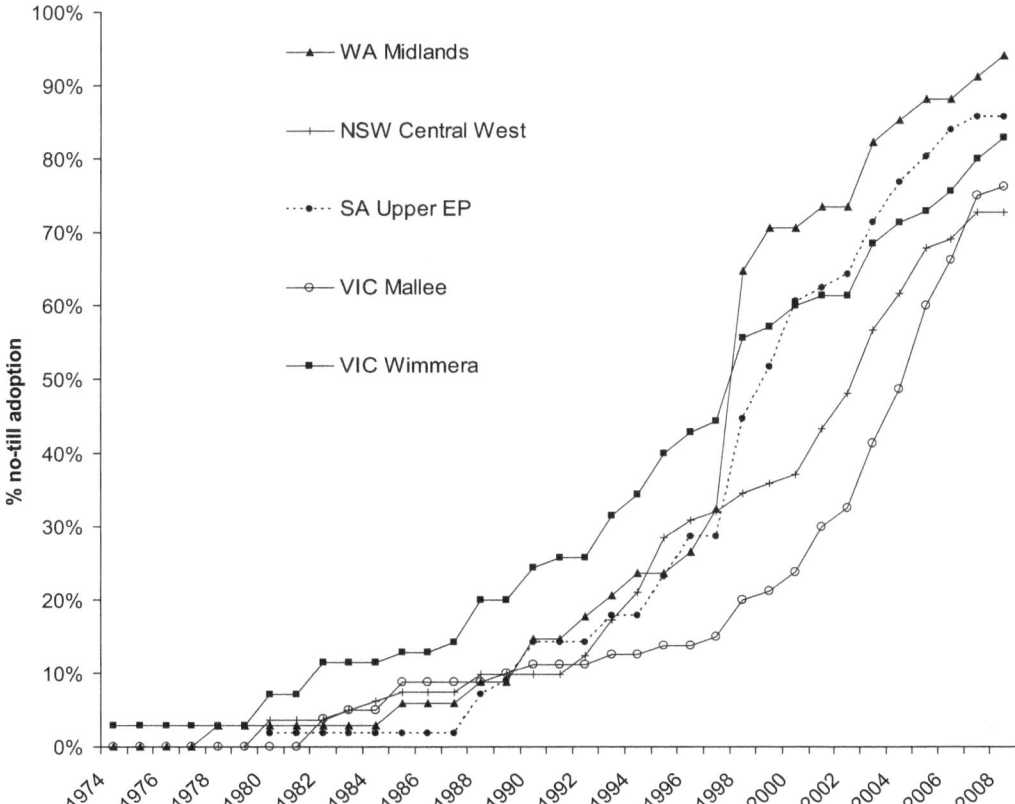

Figure 6.1. Cumulative proportion of current growers from selected Australian grain growing regions adopting use of no-till. (Based on reported year of first no-till use, from Llewellyn and D'Emden 2010.)

Factors influencing no-till adoption

In 2003, a study involving interviews with 384 farmers was conducted to identify important factors influencing the adoption of no-till in selected southern Australian cropping regions (see D'Emden *et al.* 2008). When using simple open-ended questions, the most common reasons stated for adopting no-till were soil conservation/erosion prevention, seeding timeliness and moisture conservation (D'Emden and Llewellyn 2006). The most common reasons stated for non-adoption was high machinery costs.

However, it was only when more sophisticated analysis was applied that the factors that differentiate adopters and non-adopters were identified. In this case, logit regressions (D'Emden *et al.* 2008) and duration analysis (D'Emden *et al.* 2006) were used. Duration analysis allowed the influence of variables that change over time such as prices, rainfall and farm size to be taken into account. For example, the fall in the price of the herbicide glyphosate (relative to diesel) was identified as a significant factor driving the timing of no-till adoption. This is an economic factor that is clearly out of the control of typical R, D & E projects.

The study also found that the likelihood of growers trying no-till for the first time rose after drier than average years or droughts. This suggests that droughts and dry early-season conditions can be a learning-based trigger for adoption as some of the benefits of no-till such as moisture conservation and the ability to seed on less rain become more apparent.

The results also highlighted the information-intensive nature of the complex change to no-till farming systems as reflected in the importance of availability of local information and use of farm-specific advice from agronomy consultants. The presence of nearby growers with experience with no-till was particularly important in explaining adoption. Where growers had had more time to observe no-till being practiced in their local district, the likelihood of adoption was higher.

Targeting extension for no-till adoption

Perceptions that were significantly associated with a greater likelihood of adoption were production-oriented. These included the perceived effectiveness of important pre-crop emergence herbicides under a no-till system (non-adopters were less likely to believe that key pre-crop emergence herbicides were as effective in no-till systems). Perceptions relating to seeding timeliness and moisture conservation were also important.

In terms of identifying priority learning opportunities, perceptions that are not significantly associated with adoption are often of as much interest as those that are. Although soil conservation and erosion prevention were the most common reasons for adoption stated by no-till users, perceptions relating to the erosion prevention benefits of no-till did not differentiate between adopters and non-adopters. Essentially, non-adopters and adopters were equally aware of the extent to which no-till adoption would substantially reduce erosion risk. These results suggest that continuing efforts to demonstrate erosion prevention benefits of no-till to non-adopters was not likely to be effective. The perceived effectiveness of pre-emergence herbicides for early-season weed control under no-till systems, and water-use efficiency in no-till seeding systems were likely to have greater potential as targets for information and extension.

As suggested by Figure 6.1, there is more no-till experience, observation and information available in local regions now than in 2003 (Llewellyn and D'Emden 2010). Yet adoption remains relatively low in some districts and appears set to remain relatively low over at least the medium term. If high levels of no-till use and reduced erosion risk in these regions are reached, then it will probably involve a different set of drivers and constraints to those discussed above. In regions where no-till may have lower relative advantage, reliance on information-based approaches is likely to have low effectiveness. New and innovative approaches to support extensive adoption are probably required. This is the subject of ongoing research.

When extension alone may not be the best option

In 2004, a study of durum wheat adoption by Western Australian grain growers was conducted (Nguyen *et al.* 2007), motivated by industry interest in expanding production to a level that would enable greater export market opportunities. Despite considerable efforts at development and extension, the durum wheat industry in Western Australia has remained small. The study used data from a survey of 60 Western Australian grain growers who had expressed an interest in durum wheat. Perceptions of the yield potential of durum relative to bread wheat were most influential in the decision to adopt durum. While there appeared to be some potential for extension to influence perceptions and advance learning relating to disease resistance among non-adopters, the most salient result was that growers who had recently grown durum, on average, had no higher expectations of the relative gross margins (and yield) of durum than non-adopters. Critically, growing of durum in the past was not a significant predictor of future use.

The results showed that adopters were not generating clear positive information about the innovation around key drivers of the adoption decision. This meant that programs using information-based diffusion methods aimed primarily at increasing the number of growers planting some durum by observing the experience of others were likely to be inappropriate and ineffective for reaching production targets. More emphasis on research (e.g. new varieties) and/or development (e.g. focused development of agronomic practices) to increase relative advantage appeared necessary before further investment in broader extension activities.

Steps to identifying adoption drivers and extension targets

The examples above followed similar steps used to identify a set of factors that may influence the relative advantage of a specific practice change, including the learning aspects required; survey adopters and potential adopters to collect quantitative data on those factors; and establish a subset of factors that explains a high proportion of adoption decisions. In the two examples above, a relatively small subset of explanatory variables was used to correctly categorise over 80% of respondents as adopters or non-adopters.

The method is largely focused on the decision to first use the practice rather than the decision concerning the extent of use of the practice on the farm (see Kaine *et al.*, this volume). Extensive practice change at a landscape scale obviously requires both extensive decisions to adopt the practice and the extensive use of the practice on the land managed by those adopters. These can be two very different decision processes (Abadi Ghadim *et al.* 2005). Extent of use once adopted is likely to be dominated by the on-farm experience rather than extension sources external to the farm. As Abadi Ghadim *et al.* (2005) demonstrated, there can be increases in skill through learning-by-doing once a practice is first used. This is not considered in any detail here. The focus in this paper is on the process leading to the decision to first use a new practice.

The steps used to identify opportunities to target information and learning were as follows:

(1)   Use the experience of a small number of adopters and non-adopters and other experts to develop a set of factors that may influence the likely value of the innovation for an individual grower on their farm relative to the original practice. This would mainly consist of variables needed to evaluate the innovation, including the economic and information variables contributing to expected relative advantage, such as riskiness, profitability and costs.

(2)   Collect data from a representative sample of growers that includes some adopters, using a fully-specified questionnaire that quantifies as many variables as practicable from step 1, including farmers' current perceptions of the factors that may influence overall perceived value of the innovation. Ideally, quantification of key perceptions would also include some measure of or proxy for the farmer's certainty or uncertainty about the information.

(3)   Identify variables where there are possible inconsistencies between farmers' current knowledge and field experience and/or established research knowledge, including differences between perceptions of growers who have had experience using the innovation and those who have not.

(4)   Conduct regression-based analyses to identify variables that have the biggest influence on the likelihood of adoption.

(5)   Identify the perceptions from step 3 where opportunities for learning through extension-related activities may exist that are also influential in the adoption decision in step 4.

An example where a similar set of steps to those described above was followed is in Llewellyn *et al.* (2005). Significant factors in the decision of grain growers to adopt integrated weed management (Llewellyn *et al.* 2007) were the main focus with perceptions and adoption decisions evaluated over a one-year period.

The study found that:

• targeting certain perceptions in extension activities led to measured changes in learning and subsequent changes in adoption decision-making;
• some perceptions could not be influenced because local non-adopters already had similar expectations and perceptions as local adopters (and researchers);
• for some perceptions, farmer expectations differed from researcher expectations, but farmer perceptions were not influenced by researcher opinion; and
• some common misperceptions held by farmers that were not identified as being significant in the adoption decision were substantially influenced by established research knowledge presented in the extension activity, but this did not lead to changes in adoption decisions.

Conclusion

Information and learning is central to the adoption process that leads to practice change. Factors associated with learning can explain much about the rate of adoption and whether adoption is likely to occur at a given point in time. By considering factors likely to influence the relative advantage of adopting a new practice, including learning-related variables and perceptions of key factors, it is possible to explain a high proportion of adoption decisions for a particular practice.

Where potential adopters hold perceptions that differ from those held by comparable users of a practice, there is the potential for learning. Where those perceptions are also associated with a greater likelihood of adoption, then extension that promotes that learning is likely to lead to more rapid and better-informed adoption decisions. Identifying where further extension and information is unlikely to have any impact, such as factors that cannot be influenced and/or are not influential in the adoption decision, can be just as beneficial by avoiding ineffective use of finite extension resources.

For most projects, information and learning is the primary tool available to affect practice change. The extent to which the steps used above can be followed is dependent on project aims, project size and available resources. However, the principle of 1) collecting information on the factors related to the relative advantage of the innovation, 2) assessing which factors most likely influence the adoption decision, and 3) evaluating which of these factors can be effectively influenced, is widely applicable. The approach provides an opportunity to prioritise the R&D needed to generate the most valuable information, and target extension investment to where it can best improve and accelerate adoption decisions.

References

Abadi Ghadim AK, Pannell DJ and Burton MP (2005) Risk, uncertainty and learning in adoption of a crop innovation. *Agricultural Economics* **33**(1), 1–9.

Australian Government (2008) *Caring for our Country: Business Plan 2009–2010.* http://www.nrm.gov.au/publications/books/pubs/business-plan.pdf.

D'Emden FH and Llewellyn RS (2006) No-till adoption decisions in southern Australian cropping and the role of weed management. *Australian Journal of Experimental Agriculture* **46**(4), 563–569.

D'Emden FH, Llewellyn RS and Burton MP (2006) Adoption of conservation tillage in Australian cropping regions: an application of duration analysis. *Technological Forecasting and Social Change* **73**(6), 630–647.

D'Emden FH, Llewellyn RS and Burton MP (2008) Factors influencing adoption of conservation tillage in Australian cropping regions. *Australian Journal of Agriculture and Resource Economics* **52**(2), 169–182.

Feder G and Umali DL (1993) The adoption of agricultural innovations: a review. *Technological Forecasting and Social Change* **43**(3–4), 215–239.

Fischer AJ, Arnold AJ and Gibbs M (1996) Information and the speed of innovation adoption. *American Journal of Agricultural Economics* **78**(4), 1073–1081.

Leathers HD and Smale M (1991) A Bayesian approach to explaining sequential adoption of components of a technological package. *American Journal of Agricultural Economics* **73**(3), 734–742.

Lindner RK (1987) Adoption and diffusion of technology: an overview. In *Technological Change in Postharvest Handling and Transportation of Grains in the Humid Tropics.* (Eds BR Champ, E Highly and JV Remenyi) No. 19. pp. 144–151. Australian Centre for International Agricultural Research: Bangkok, Thailand.

Lindner RK, Pardey PG and Jarrett FG (1982) Distance to information source and the time lag to early adoption of trace element fertilisers. *Australian Journal of Agricultural Economics* **26**(2), 98–113.

Llewellyn RS (2007) Information quality and effectiveness for more rapid adoption decisions by farmers. *Field Crops Research* **104**(1–3), 148–156.

Llewellyn RS and D'Emden FH (2010) 'Adoption of no-till cropping practices in Australian grain growing regions'. GRDC-CSIRO Report. www.grdc.com.au/notill_adoption.

Llewellyn RS, Lindner RK, Pannell DJ and Powles SB (2005) Targeting key perceptions when planning and evaluating extension. *Australian Journal of Experimental Agriculture* **45**(12), 1627–1633.

Llewellyn RS, Lindner RK, Pannell DJ and Powles SB (2007) Herbicide resistance and the adoption of integrated weed management by Western Australian grain growers. *Agricultural Economics* **36**(1), 123–130.

Marra MC, Hubbell BJ and Carlson GA (2001) Information quality, technology depreciation, and Bt cotton adoption in the southeast. *Journal of Agricultural and Resource Economics* **26**(1), 158–175.

Marsh SP, Pannell DJ and Lindner RK (2000) The impact of agricultural extension on adoption and diffusion of lupins as a new crop in Western Australia. *Australian Journal of Experimental Agriculture* **40**(4), 571–583.

Nguyen VH, Llewellyn RS and Miyan MS (2007) Explaining adoption of durum wheat in Western Australia. *Australian Agribusiness Review* **15**, 14–26.

Pannell DJ, Marshall GR, Barr N, Curtis A, Vanclay F and Wilkinson R (2006) Understanding and promoting adoption of conservation practices by rural landholders. *Australian Journal of Experimental Agriculture* **46**(11), 1407–1424.

7

Enabling change in family farm businesses

Amabel Fulton and Frank Vanclay

Summary

The argument of this chapter is that a deep understanding of the family farm business, including its components and their interactions, is critical to designing effective mechanisms for enabling change in Australian agriculture. The process of enabling change, i.e. extension, needs to address the *needs* of the *right* people, at the *right* time. This chapter presents a model that conceives family farm business behaviour as the outcome of the interactions between the family, the natural resources which are farmed, and the farm business, all within a broader social, economic and political context. A service brokering model which incorporates this approach is proposed as an alternative to the current 'product-push' model of past and current extension practice.

Introduction

Family farming has been and still is the backbone of Australian agriculture. Since settlement, family farm businesses have prevailed in the face of ongoing and repeated threats to their survival: fire, drought, environmental degradation, declining terms of trade, debt, the technological treadmill, declining social status, reduced services in rural communities, and market domination of multi-national processors, distributors and retailers (Vanclay and Lawrence 1995). According to many theorists from a range of disciplines, however, family farm businesses exposed to these kinds of forces are unlikely to survive. Yet despite this prognosis, most Australian farm businesses – large and small – continue to be owned and managed by families. While it is difficult to obtain precise figures, and they vary by commodity, it is likely that family farming exceeds 90% of all farms (Gray and Lawrence 2001; Vanclay 2003).

Australia's family farm businesses operate within an advanced industrialised agricultural system. The industrialisation of agriculture over the last century has led to massive changes in the social and physical structure of agriculture. This process has included mechanisation (e.g. tractors, harvesters and irrigation), chemical farming (e.g. fertilisers and pesticides), and food processing (canning and freezing). These activities have allowed the penetration of industrial and financial capital into the agricultural production system through, for example, the provision of credit, the supply of farm inputs, and vertical integration. As such, farm businesses are a small part of a larger food system which encompasses all aspects of food production, from the supply of agricultural inputs, through on-farm production, to food processing, distribution and consumption. The system is influenced by the relationships between its components and by the physical farm environment, farm policies at state and federal levels, international food trade, credit and financial markets (Bowler 1992).

The industrialisation of Australia's agriculture has made it more productive and more efficient, but also more expensive and more environmentally damaging. While the gross value of farm production has been increasing, the terms of trade for Australian family farm businesses have declined. In addition, there has been a continuing decline in the number of farms, perhaps by as many as 2000 per year (Vanclay 2003; see also Garnaut and Lim-Applegate 1998), and in the number of people employed on farms. Over time, there has been a shift in income sources from the farm being the only source of income to a situation where much income is earned off-farm. Thus Australia's family farm businesses have experienced significant changes. These changes have presented opportunities and challenges for family farm businesses seeking to continue in the long term.

Extension has been an agent in the delivery of many of these changes, as well as being a source of assistance to prevail. Extension itself has undergone change in the face of neoliberalism, initially from providing one-on-one technical services to farmers at a local level free of charge, to providing a reduced service through group extension, to providing limited services on a fee-for-service arrangement akin to commercial agriconsultants, to being largely dismantled as a government activity in some Australian states. Nevertheless, extension services continue to be provided by a new range of deliverers, such as commercial agribusiness, agricultural consultants of various types, and by a range of non-government organisations (NGOs).

Enabling change in rural Australia

Over time, governments have continued to support family farm businesses to manage change to ensure the long-term sustainability of Australia's farms and of the natural resources they manage. One of the key mechanisms used has been extension. Extension is 'the process of enabling change in individuals, communities and industries involved in the primary industry sector and with natural resource management' (SELN 2006, p. 4). Historically, extension was a service delivered by staff of state governments to support on-farm production and to accelerate adoption of the latest research and development. This support, which delivered private benefits to family farm businesses, was considered to deliver public benefits in terms of economic growth and food security. During the last decade or so, the funding mechanism for extension has changed, and natural resource management has become a major element of the extension message and rationale (Vanclay and Lawrence 1995).

Over the years, family farm businesses have had access to a plethora of support to assist them to manage change. The effectiveness of this support, in terms of achieving the goals of the investors (i.e. government, relevant NRM body, etc.) as well as the goals of the family farm businesses themselves, has been questioned. This chapter brings together ideas about the family farm business and extension (as delivered by government, NGOs, and commercial consultants) to explore some of the limitations of extension delivery in Australia, particularly in relation to its ability to address the needs of the family farm businesses it seeks to support.

Understanding family farm businesses

Defining the family farm is difficult because there are many different aspects to the family farm business, and definitions tend to be constructed for theoretical purposes. For example, some emphasise the role of family members in the farm work; others focus on the ownership of the farm by the family; while others examine the nature of the relationships between the farm family with non-family organisations. Gasson and Errington (1993) identified an ideal type of family farming that comprised five characteristics:

- a combination of ownership and management
- business principals (i.e. owners) being related by kinship or marriage
- family members providing capital and labour
- the family living on the farm
- intergenerational transfer of assets and control.

Their approach was distinct from other definitions in that while family members were required to provide capital and labour, they were not required to provide *all* of the capital and *all* of the labour *all* of the time. In addition, the family farm is considered a business, a productive unit providing a return on land, capital and labour to its owner-managers.

Gasson and Errington's definition, however, gives little prominence to the farm (as a biophysical, environmental entity) as a key component of the family farm business. Rather it prioritises the business and family components. As such, it ignores the usually strong relationship between the family and the land, as well as the link between the economic success of the business with the farm's natural resources and biophysical constraints.

The family farm business has been and still is the predominant structural unit of Australian agriculture. Of the 140 700 farm businesses in Australia in 1997, 98% were run by non-corporate owner-managers (Garnaut and Lim-Applegate 1998). Of these, 81% were run by two or more family business partners; at least 85% of the labour used in these businesses was provided by members of the farm business; 73% of families lived on the farm (Garnaut and Lim-Applegate 1998); and in the majority of cases there had been intergenerational transfer of the farm business to family members, and an ongoing desire to continue that tradition. It is likely that these figures would still be generally indicative of the situation now a decade later, even though the total number of farmers will have declined further.

Enabling change in family farm businesses

Our primary argument is that a deep understanding of the dominant structure of Australian agriculture, the family farm business, is central to enabling change in rural Australia. The family farm business is the total of the elements and interactions between the family, the farm and the business. In contrast to most commentators who fall into the habit of regarding 'farm' and 'farmer' as singular, as sole operator, we argue that the family farm business is an organisation (albeit a social organisation), and the workings of this organisation need to be well understood if efforts to enable change in rural Australia are to be effective. A deep understanding of the family farm business – its components and its interactions – is critical to designing extension which addresses the *needs* of the *right* people at the *right* time. A service brokering model – which incorporates this approach – is proposed as an alternative to the current product-push model of extension in Australia. A service broker is defined as an individual or organisation that takes an active and purposeful role in identifying service needs. A service broker considers the whole suite of present and potential product and service opportunities and actively matches needs to products and services, acting in the best interests of the client. Service brokers will be most effective in their role when they understand the family nature of the farm business.

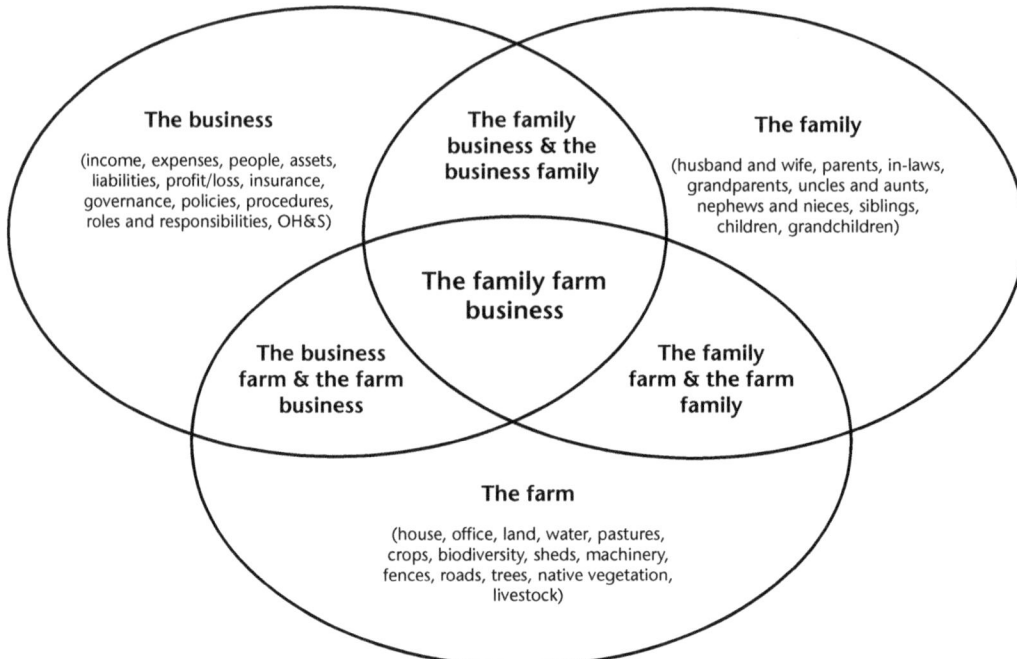

Figure 7.1.   A conceptual framework for understanding the family farm business.

(Source: Fulton 2009.)

A conceptual framework for understanding the family farm business

Family farming can be considered as the interaction between the family, the land which is farmed, and the farm business (see Figure 7.1). As such, any change to the nature of any of these components potentially has an impact on the nature of family farming, and on the other components of the family farm business.

From this conceptual model of the family farm business, a number of distinctions can be made:

- the family business
- the business family
- the family farm
- the farm family
- the business farm
- the farm business
- the family business farm
- the business family farm
- the family farm business
- the farm family business
- the business farm family
- the farm business family.

This list demonstrates the variety of forms the family farm business can take. Family farm businesses are all of these at different times, in different situations. Unlike Gasson and Errington (1993), this framework also demonstrates the importance of the physical farm as a component of the family farm business.

Interactions between the components of the family farm business

This section examines the interactions between the different components of the family farm business by considering how a change in one component affects the others. We then explore how the whole interacts with the social, political, environmental, technological and economic forces acting on it.

Changes within the family affect the business

A change to the farm family, such as the arrival of children, may reduce the availability of parents for farm work. This impacts on the farming business as decisions need to be made as to: (a) whether the level of labour provided to the farming operation should be maintained (and one partner may need to work longer hours or employ additional labour); (b) whether changes should be made to the farming operation to reduce its labour requirements and/or increase profitability; and (c) whether the role of the parents may need to change to allow for continued contribution to the farm business while also providing care for the new family members.

Changes to the family affect the condition of the farm

Changes to the farm family and subsequent operation of the farm business will in turn have consequences for the land which is being farmed. A reduction in family labour availability may mean that particular farm activities considered 'non-essential' are put on hold for some years. What is considered 'non-essential' will depend on the family in question, but could include activities with a low return or of a long-term nature. These may be activities such as tree planting, fencing-off riparian zones, or re-fencing the property according to land suitability. In addition, a family may look at ways to cut costs, such as by reducing fertiliser inputs, or reducing stock numbers to eliminate the need for hand-feeding during poor seasons. Any of these changes will affect the condition of the farm's natural resource base.

Changes to the farm family can indirectly affect the natural resources of the property through changes to the farming operation. The presence of a successor, or even the possibility of a successor, can provide motivation for long-term decision making – that is, decision making beyond the life of the current parents and even of their offspring – to ensure the business and the land are in the appropriate condition to continue to support the farm family for generations to come. Decisions that in other circumstances might seem economically irrational due to the long timescale for return on investment – such as the planting of trees for shelter or biodiversity – are more easily taken within such an intergenerational mind-set.

Changes to the farm's natural resources affect the business and then the family

Changes to the condition of the farm's natural resources affect the farm business and subsequently the farm family. Examples of such changes include reduced productivity through soil erosion or fertility decline, salinity, declining water quality, and declining biodiversity. These phenomena will affect each family and its farm business differently, depending on the nature of each family and on the nature of its farming business. The impact will be buffered to a greater or lesser extent by factors such as debt levels and the dependence of the family farm business on the income from the degraded areas. As discussed previously, the impact on the farm business will have consequences for the farm family as a social and economic unit.

Changes to the farm's natural resources affect the family and then the business

While the indirect impact of natural resource decline on the farm family is evident, such changes also directly affect the farm family, which in turn affects the farm business and the condition of the natural resources. Some farm family members may, for example, value the land's capacity to be productive very highly. The decline in the land's ability to produce may significantly affect their sense of self-worth with consequent impacts on their mental health and on other farm family members. This, in turn, can affect the running of the farm business, as morale and optimism declines, and the family farm business can get caught in a downward spiral of economic, social and environmental decline. Alternatively, such a farm family member's passion for a productive property may inspire them to take innovative approaches to ensuring the long-term viability of their business and property.

External forces and the family farm business

The discussion thus far has demonstrated the impact of family changes on the farm business and on the condition of the natural resource base, and how a change in any one of these three components has consequences for the other components. This discussion so far has been limited to internal factors – those factors within the farm family, within the farm business, or within the natural resources. External factors can also affect the natural resources, the farm business or the farm family; with subsequent implications for the interacting components of the family farm business (see Figure 7.2).

The impact of external forces on the business (and thus on the family and the farm)

Examples of external factors may be drought, declining commodity prices, changing societal norms, and changing government policy. Each of these factors directly affects one or more components of the family farm business, with flow-on consequences for the other components. Drought and/or declining commodity prices, for example, may mean that the income from the farm becomes insufficient to support the family unit. Members of the farm family may respond by altering their own activities (such as by seeking off-farm work or reducing living expenses) or by altering the farming activities (such as reducing farm expenses or increasing productivity). Drought and changes to farming activities will also affect the condition of the natural resources, with subsequent consequences for long-term environmental sustainability.

Impact of external forces on the family (and thus on the business and the farm)

As discussed previously, drought can also affect the social health of family members, with flow-on effects for the business and natural resource management. Declining economic performance also affects family member morale, with consequences for farm management and natural resources. Societal changes also act directly through the farm family. An example of societal change is women seeking to improve their status through seeking professional employment off-farm, or by taking a greater role in the management of the business. This in turn has consequences for the farm business and the farm, whether through improved off-farm or on-farm returns, or through the injection of new enthusiasm and skills into the management of the business and farm.

This discussion has illustrated the impact of internal and external factors on the family farm business. A complex and interactive system is apparent, with multiple components and

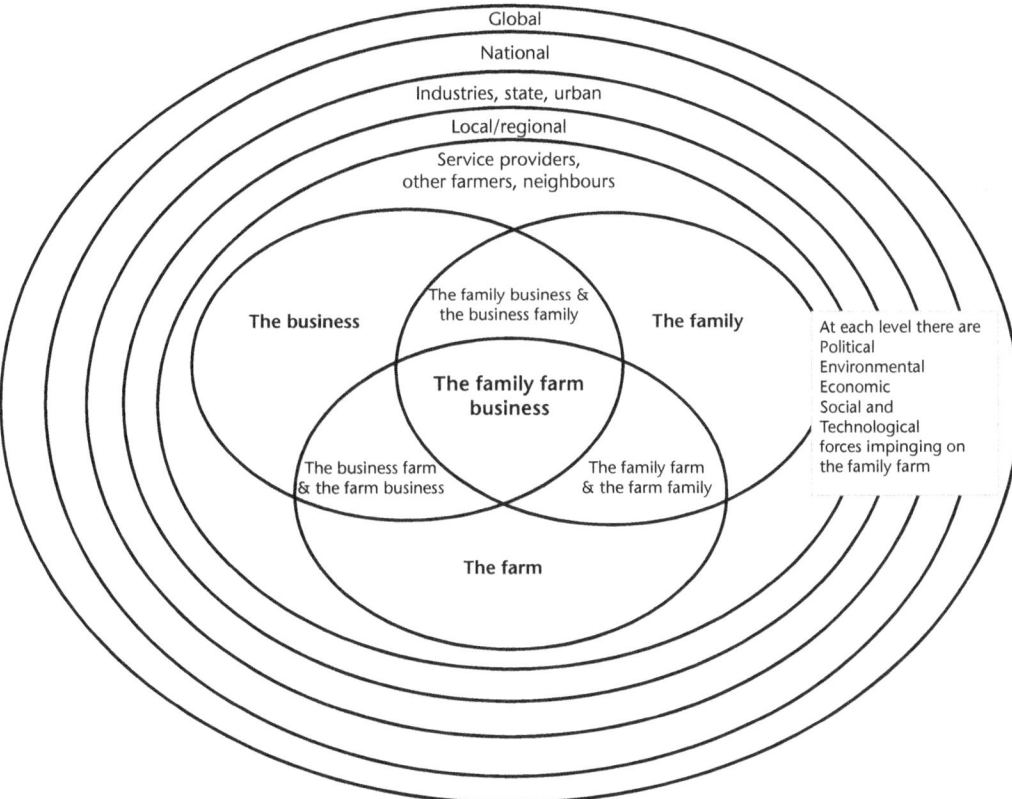

Figure 7.2.   The interactions between the family farm business and the broader environment.

(Source: Fulton 2009.)

factors playing a part in determining the direction of the family farm business. This system can be further elucidated through consideration of the sub-components of the family farm business, and of the wider social and economic context within which it is operating.

Impact of external forces on the farm (and thus on the business and the family)

Examples of external forces impacting on the natural resources base of the farm include drought, climate change, road developments (cutting through farm land); water storage developments (inundating and/or resuming farms); and new regional infrastructure developments (e.g. pulp mills or mines). Drought can lead to soil erosion, pasture degradation, weed proliferation and tree decline. Climate change can alter the microclimate of the property. Major external infrastructure developments can have multiple impacts, depending on their specific nature, the most obvious one being the loss of access to natural resources. These changes can have positive and negative effects on the business, on the psyche of the family, and on the ability of the farm to provide sufficient income to continue to support the family's needs.

Enabling change in family farm businesses: a new perspective

The discussion so far has demonstrated the complexity of family farm businesses and the interrelatedness of their components. Family farm businesses are working in an environment of constant change – whether that be internal from changes to the family circumstances, or external as a result of government policies or climatic change. Many of these changes are occurring at the same time. For each family, the specific changes being dealt with at any one point in time will be a function of the nature of the family, the nature of the farm's biophysical resources and their location, and the nature of the farming business. The changes being dealt with are dynamic – changing as the family, the farm biophysical resources and the business changes and adapts. The changes any particular family farm business are experiencing will be specific to them and, while some of the forces will be similar to those experienced by other family farm businesses, the nature and extent of the impact of each force will differ for each family farm business.

Just as Maslow's hierarchy of needs illustrates that an individual's base needs must be met before higher needs become a priority, so too must a family farm business's specific needs at a particular time be addressed before other issues become of significant concern. This means that before family farm businesses will be interested in listening to the latest extension priorities of government or industry, they will be seeking to address their own needs first. Of course, the member of the family farm business who has greatest influence in the particular area of need will normally be the one who will have most influence on the actions that are taken to address that need.

To enable change, then – as extension is intended to do – efforts need to focus on addressing the specific needs of individual family farm businesses: at the right time (i.e. when it is a priority for the family) and with the right people (i.e. the appropriate member(s) of the family farm business for that particular issue). Extension designers and providers need to understand and address the specific needs of each family farm business at the right time with the right people.

Product-push versus needs-based extension

Family farm businesses have access to a wide range of services and often public funding to assist them in the process of managing change, including:

- publicly funded, publicly delivered extension (e.g. state government farm advisers)
- publicly funded, privately delivered extension (e.g. property management planning delivered by private consultants engaged by the regional NRM bodies funded by the federal government)
- privately funded, privately delivered extension (e.g. business consultants paid for fully by the client)
- jointly funded extension delivered by both the private sector and the public sector, for example:
  - Landcare (where federal government investment is matched by in-kind support from the private sector)
  - FarmBis (where the cost of training is partly paid for by participants, and partly by a joint state and federal government subsidy)
  - Research and Development Corporations (where levies from farm production are matched by the federal government and allocated to industry research, development and extension).

From the family farm business perspective, there is a continuously changing plethora of services on offer, usually delivered by a person on a short-term contract. The family farm business may or may not be aware of all the services on offer. The short-term employee's role is to engage farmers (not necessarily family farm businesses) in using the products and services offered through the employee's project. These products and services may – or as usually is the

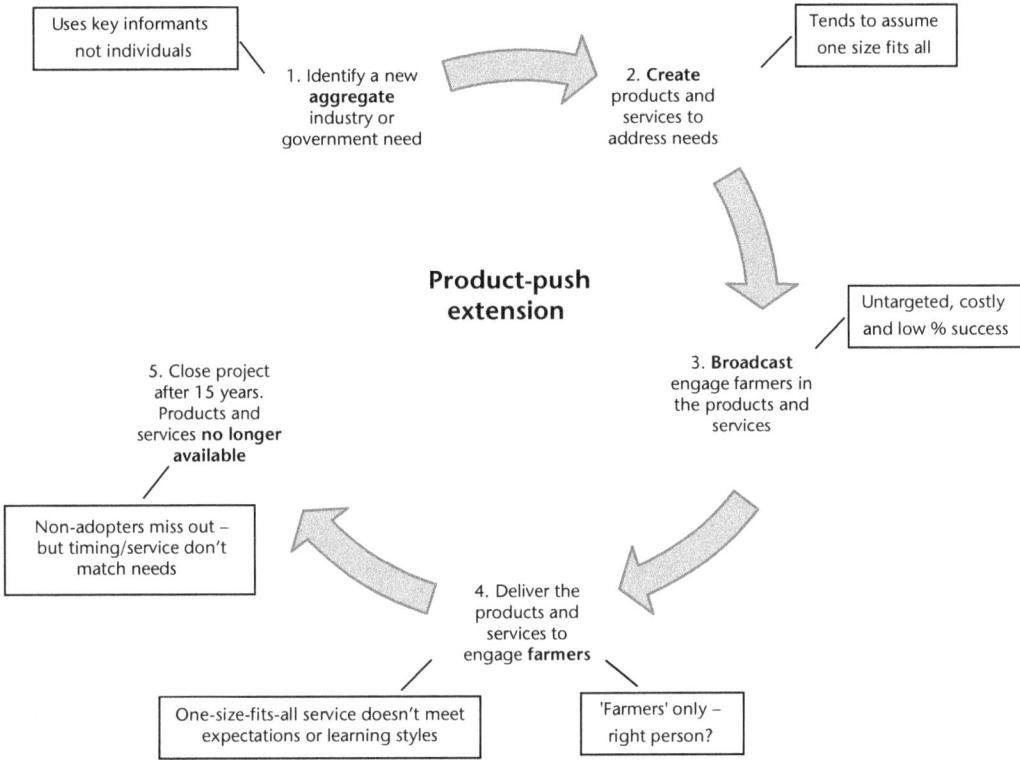

Figure 7.3.   The processes and limitations of the product-push model of extension.

(Source: Fulton 2009.)

case, may not – meet the needs of the family farm business at that time. More often than not, the short-term employee has a low level of understanding of the needs of family farm businesses in general, or the specific needs of the family farm business the employee is targeting. With each service provider focused on delivering their project, there is no single service provider who has an holistic understanding of the needs of the family farm business, or businesses in any particular industry or region. This product-push model of extension is shown in Figure 7.3.

The product-push model is typically used by the Research and Development Corporations and government to design and deliver on industry and government priorities. The key point to note about this model is that the needs are formulated at an aggregate level at the outset, with little customisation taking place at any stage in the process.

The major limitations of the existing service delivery models are:

(1)   The products and services provided are largely determined by the funder or the extension organisation, with little consultation with the end-users until the time of product delivery
(2)   The product or service is delivered by someone with little knowledge or understanding of the complexity of the nature and functioning of family farm businesses, or the specific family farm businesses being targeted
(3)   The 'farmer' is typically considered to be an individual, rather than being part of an organisation
(4)   The 'farmer' is typically considered to be the senior male, rather than other members of the family farm business

(5) Services are generically targeted to 'the farmer', rather than to the appropriate member(s) of the family farm business
(6) Services are delivered without consideration of the specific needs of those other than the senior male
(7) Services are delivered without consideration of the interactions (and associated impacts) between the different components of the family farm business and the external environment
(8) Services are provided as and when it suits the service provider, rather than at a time which best suits the family farm business.

These limitations lead to inefficiencies and ineffectiveness in the delivery of individual extension projects as short-term employees often only work out what *should* have been done when the project finishes. In addition, there is a huge duplication of services at the regional level as each industry or sector seeks to engage with the same family farm businesses on similar issues.

Services which address these limitations are more successful in enabling change (Kilpatrick *et al.* 2002, 2006). Such services are provided with a thorough understanding of the needs of family farm businesses, in the context of their external operating environment at the time. This ensures a needs-driven approach to enabling change, rather than the 'product-push' approach to change. In particular, the process of 'brokering' – identifying family farm business needs and then matching services to address them – provides a systematic framework for enabling change with family farm businesses. Such an approach can ensure that the right members of the family farm businesses are receiving the right services at the right time. This model of needs-based extension is presented in Figure 7.4.

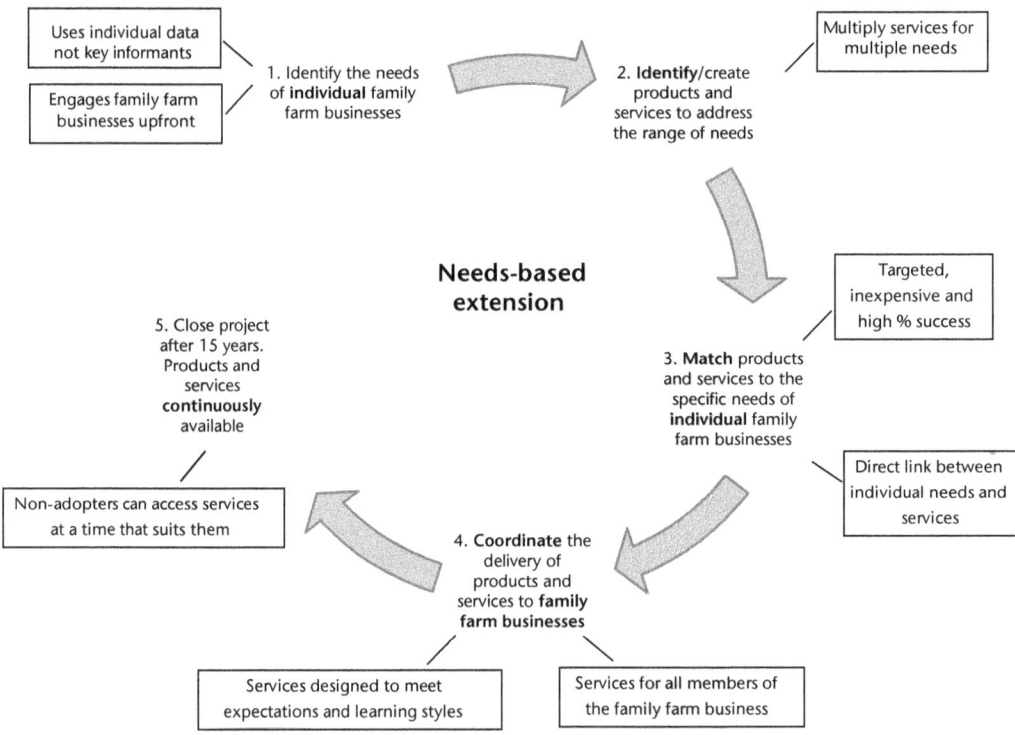

Figure 7.4. The processes of needs-based extension.

(Source: Fulton 2009.)

The key difference between the needs-based model (Figure 7.4) and the product-push model (Figure 7.3) is that in the needs-based model, the products and services are designed to address the specific needs of the family farm businesses involved, while still delivering on the industry or government needs. Engagement takes place as the first stage of the process and is maintained throughout the project. The products and services may or may not be those delivered within the scope of the core project – they are likely to include additional products and services available from other sources, which meet the specific needs of the family farm businesses.

Ideally, the products and services created have a built-in legacy plan to ensure their continued availability to family farm businesses in the future. For example, particular processes or information packages could become standard services within the private sector, delivered as and when particular family farm businesses require them.

This model addresses the eight limitations listed earlier in the following manner:

(1) The products and services provided are largely determined by the family farm businesses through full consultation with end-users at the time of product development
(2) The products and services are designed by people with significant knowledge and understanding of the complexity of the nature and functioning of family farm businesses
(3) 'Farm' does not become equated with 'farmer', but rather with 'family farm business', and individuals although individually different are understood to be participants in a family farm business
(4) The word 'farmer' is used cautiously, and is applied to all members in the family farm business and not just the senior male
(5) Services are targeted to the family farm business via the appropriate member(s) of the family farm business for that area of decision making, which may well be different in different family farm businesses and for different services
(6) Services are delivered with consideration given to the specific needs of all members of the family farm business
(7) Services are designed and delivered with full consideration of the interactions and associated impacts between the different components of the family farm business and the external environment
(8) Services are provided as and when it suits the family farm business, rather than at a time which best suits the funding body or service provider.

Adoption of the principles and practices of needs-based extension by extension designers and deliverers has the potential to increase the effectiveness and efficiency of extension project delivery significantly. The increased effectiveness comes from the allocation of resources directly to the issues of concern to the family farm business, rather than allocating resources to persuading the family farm business to reconsider its needs and priorities to match those of the service provider, or to 'finding' family farm businesses whose needs match those of the extension project. Engagement is strong because the extension provider is authentically attempting to address the needs of the family farm business. Engagement is also maintained over time, leading to increased trust and credibility of the extension provider over the long term (a major factor in enabling change). By working with the appropriate member of the family farm business, the extension provider is also well placed to influence decision making within the farm unit. Ultimately, because the products and services are matched to the specific needs of each family farm business, the rate and extent of adoption is much higher than when the degree of matching is not so complete.

Across a region, the adoption of needs-based extension by a range of institutions has the potential to reduce the investment needed to support change significantly, or alternatively, to increase the efficiency of allocation of this investment just as significantly, resulting in an accelerated rate of change or adoption within the region. By adopting a needs-based approach which is fully informed of the other extension products and services available in the region, institutions can coordinate their efforts to broker their products and services across a wider audience in a more direct, timely and targeted manner.

While this discussion tends to contend that product-push extension is predominant in Australian extension, there are examples of programs where elements of the needs-based model are incorporated into the extension design. There are also examples of needs-based extension being delivered, but this is more likely to occur where the extension providers have long-term, locally based relationships with family farm businesses and, as such, are better placed to respond to the changing needs of family farm businesses in their region. There are, however, few examples of coordinated approaches to extension delivery. Until extension in Australia is able to rationalise its investments and work cooperatively, the returns to government, industry and individuals will continue to be much less than is possible. A shift in focus from product-push to needs-based coordinated extension will lead to improved sustainability of our family farm businesses, our rural industries and rural and regional Australia.

References

Bowler I (1992) The industrialization of agriculture. In *The Geography of Agriculture in Developed Market Economies*. (Ed. I Bowler) pp. 7–31. Wiley: New York.

Fulton A (2009) Enabling change in family farm businesses. PhD thesis. University of Tasmania: Hobart.

Garnaut J and Lim-Applegate H (1998) *People in Farming*. ABARE: Canberra.

Gasson R and Errington A (1993) *The Farm Family Business*. CAB International: Wallingford.

Gray I and Lawrence G (2001) *A Future for Regional Australia*. Cambridge University Press: Cambridge.

Kilpatrick S, Fulton A and Geard L (2002) 'Providing client focused education and training'. Report to Department of Agriculture, Fisheries and Forestry Australia. Centre for Research and Learning in Regional Australia, University of Tasmania: Launceston.

Kilpatrick S, Fulton A, Johns S and Weatherley J (2006) 'A responsive training market: the role of brokers'. RIRDC Report 06/110 for the Cooperative Venture for Capacity Building. Rural Industries Research and Development Corporation: Canberra.

SELN (State Extension Leaders Network) (2006) 'Enabling change in rural and regional Australia: the role of extension in achieving sustainable and productive futures'. A discussion document produced by the State Extension Leaders Network August 2006. http://www.seln.org.au.

Vanclay F (2003) The impacts of deregulation and agricultural restructuring for rural Australia. *Australian Journal of Social Issues* **38**(1), 81–94.

Vanclay F and Lawrence G (1995) *The Environmental Imperative: Ecosocial Concerns for Australian Agriculture*. Central Queensland University Press: Rockhampton.

8

What 'community' means for farmer adoption of conservation practices

Graham R Marshall

Summary

'Community-based' natural resource governance has been sponsored by Australian governments since the 1980s as a way of strengthening farmers' self-reliance in adopting conservation practices. Our understanding of how this outcome may be achieved has mainly reflected thinking in the discipline of rural extension. Ideas from the theory of collective action reveal that adding a 'community' layer to natural resource management (NRM) governance means more for strengthening farmers' self-reliance than can be appreciated with rural extension thinking alone. These ideas highlight the complementary roles of 'vertical' trust and reciprocity between farmers and NRM governance structures in strengthening farmers' self-reliance in this domain. They demonstrate it is not enough that farmers trust NRM governance structures to support them in addressing the NRM problems they face. Farmers' self-reliance is strengthened under NRM governance only to the extent that this trust exists and they become motivated to reciprocate the support they receive (e.g. by complying proactively with conditions attached to this support). Potential advantages of community-based governance in providing such motivation are explored. Quantitative evidence from the author's research indicates that community-based governance can achieve significant success in exploiting these advantages and thereby can strengthen farmers' self-reliance in addressing the NRM challenges they face.

Introduction

Since the 1980s, Australian governments have sponsored participatory, decentralised approaches to natural resource management (NRM). This style of governance, often described as 'community based', has been chosen as a means for supporting farmers' self-reliance in adopting conservation practices. I argue in this chapter that too much emphasis has been placed on 'rural extension thinking' as a basis for understanding how community-based governance may strengthen this self-reliance. I argue further that this has led the Australian NRM policy community to think too narrowly about the potential of this approach. Drawing on evidence from my research, I propose that insights from the theory of collective action can broaden this understanding and help us design and implement community-based arrangements more capable of strengthening farmers' self-reliance in adopting conservation practices.

It needs to be recognised at the outset, however, that community-based NRM governance cannot work miracles. It is unrealistic to expect community-based arrangements to strengthen this self-reliance when the conservation practices on offer are contrary to farmers' interests and thus not in the vicinity of being 'adoptable'.

Australian policy supporting community self-reliance in NRM

In most developed nations until the late 1970s, people concerned about degradation of natural resources tended to expect governments to intervene. By the 1980s, however, citizens in many nations had lost considerable faith in their governments' abilities to respond to their concerns and were demanding to participate more directly in this domain. The National Conservation Strategy for Australia, prepared in 1983, emphasised the need for rural communities to participate in the planning and implementation of conservation initiatives. This emphasis was reinforced by a vanguard of soil conservationists, extension agents and farmers who were adapting emerging theories of rural development (e.g. Chambers 1983; Esman and Uphoff 1984) to a developed economy. These theories emphasised:

- self-help supported by change agents
- human resource development rather than technology transfer
- public participation
- cooperative efforts at the local community scale (Curtis 1998; Curtis et al. 2008).

The National Landcare Program (NLP), launched in 1989, is an early landmark in the Australian turn towards community-based NRM. The NLP's original emphasis was on catalysing local activity by supporting the formation and facilitation of Landcare groups, education and awareness-raising activities, and demonstration sites. The Commonwealth Department of Primary Industries and Energy stated at the time that it was 'trying to encourage a process of self-help … some day the local community has to pick up all this' (House of Representatives Standing Committee on Environment Recreation and the Arts 1989, p. 72).

Adoption by Australian governments of the concept of 'integrated catchment management' (ICM) during the 1980s and 1990s consolidated this move towards a community-based approach, although the catchments delineated for ICM programs were normally much larger than the local landscapes around which Landcare groups had formed. Given fears that 'a regulatory approach to ICM could focus farmers' energies on resisting interference by bureaucrats rather than on improved land management' (Hollick 1992, p. 51), ICM committees were expected to achieve voluntary cooperation from those they depended on for implementation of their strategies.

The focus of the NLP on fostering local self-help made state and territory governments particularly interested in Landcare groups as vehicles for implementing ICM programs with modest additional budget outlays. For instance, Cunningham (1988, p. 43), then the Chief of Services for the NSW Soil Conservation Service, argued that 'in today's economic climate where governments are faced with escalating non-discretionary expenditure, it is essential that notions of self-help be promoted to achieve catchment management. No longer can the Government foot the bill for catchment protection'.

In 1997, the Commonwealth established the Natural Heritage Trust (NHT) which differed from the NLP by focusing its investments on on-ground implementation and by channelling them principally via catchment or regional-level ICM organisations. The NHT program was presented as a framework for 'partnerships' between communities, industry and government. Concerns regarding the accountability of regional NRM organisations to investing governments (e.g. Industry Commission 1998) led to a tightening of partnership arrangements under the National Action Plan for Salinity and Water Quality (NAP), established in 2000, and the extension of the NHT (i.e. NHT2), announced in 2001.

The NAP and NHT2 programs became known jointly as the 'regional delivery model'. Governments viewed the partnership approach as a means of fostering 'community ownership' of natural resource problems. The aspiration was to foster among farmers and other community members a sense of shared responsibility in addressing environmental problems (Wallington *et al.* 2008). Policy documents preceding the launch of the regional delivery model referred accordingly to landholders having 'a mutual obligation, or duty, to manage and care for [natural] resources in a sound and sustainable manner' (Agriculture Fisheries and Forestry Australia 1999, p. 53), and to an objective of 'self-sustaining, proactive communities that are committed to the ecologically sustainable development of natural resources in their region' (Steering Committee 2000, p. 6). The National Natural Resource Management Capacity Building Framework stated that a 'strong feeling of ownership over the NRM planning process will increase motivation and the likelihood that the outcomes identified in the regional integrated NRM plans are achieved' (NRMMC 2002, pp. 5–6).

Governments continue to assume that devolution of appropriate NRM responsibilities to community-based regional organisations strengthens community members' self-reliance in addressing environmental problems. The Framework for Future NRM Programmes endorsed by the Natural Resource Management Ministerial Council (2006, p. 5) prior to completion of the NAP and NHT2 programs argued that 'strategic landscape-scale change is most effectively achieved where communities have a sense of ownership over planning and investment decisions, and will therefore make the investments of time, resources and better practices that are needed to achieve better NRM outcomes'.

The NSW Natural Resources Commission (2008, p. 3) stated, 'A key part of the CMA's role is to engage with their communities, gain their trust, build their ownership of the regional [Catchment Action Plan] and targets and then 'help them help themselves' by voluntarily adopting sound NRM practices and acting as stewards of the natural resource assets on their land'.[1] Launching the Outcomes 2008–2013 Statement for the Australian Government's Caring for our Country program (which superseded the NAP and NHT2 programs), the Minister for the Environment, Heritage and the Arts stated: 'One of the national priorities in Caring for our Country is community engagement and ownership and connection' (Garrett 2008). The importance of fostering landholders' self-reliance in addressing NRM problems was restated most recently in a Senate inquiry report as follows:

> It is very clear that the extent of land use change needed at the landscape scale to address the combined challenges of landscape degradation, weed and pest management, biodiversity conservation, and sustainable water use in a changing climate will continue to require a significant level of voluntary action and private investment by land managers and regional communities (Senate Standing Committee on Rural and Regional Affairs and Transport 2010 pp. 69–70).

Interpreting the role of community in NRM

The focus of the aforementioned NRM programs on fostering community self-reliance soon came to be interpreted predominantly through a rural extension lens given that many Australian NRM policy makers and practitioners received an education that encouraged them to view social issues in NRM as rural extension issues. Hence, the focus was mainly on strengthening the self-reliance of farmers in adopting conservation practices through rural extension activities. Largely due to government pressures at the time to wind back government expenditure on rural extension, these activities tended to employ group-based methods requiring less public funding than one-on-one methods. Kingwell *et al.* (2008, p. 904) observed how:

> In the late 1980s and early 1990s the widely held view, especially in government circles, was that farmers were unaware of key land degradation issues and they lacked the attitude, knowledge and skills necessary to address these issues ... Accordingly governments committed billions of dollars, mostly to the community-based approach, to raise awareness, to provide education resources, to offer skills training and to support research and community-based projects.

Early research into the community-based approach (e.g. Curtis and De Lacy 1996; Curtis *et al.* 2001; Vanclay 1992) demonstrated it had been successful according to the conventional yardsticks of rural extension. Farmers' awareness and knowledge of NRM issues had increased and their attitudes towards conservation had become more positive. Moreover, various studies (e.g. Cary *et al.* 2002; Curtis 1997; Mues *et al.* 1998) found a positive relationship between membership of Landcare groups and adoption of some conservation practices, although the direction of causation was not clearly established. Nevertheless, the reliance on group-based extension approaches for solving land degradation problems in agriculture came to be criticised from the late 1990s (e.g. Lockie and Vanclay 1997). For instance, Marsh and Pannell (2000, p. 624) stated that:

> We are dismayed that Government and funding bodies appear to believe that extension through Landcare groups will be sufficient to achieve widespread adoption of conservation practices. In particular, we are concerned that there is a belief that farmers can solve difficult and complex land degradation problems themselves through group-based processes, even when it is apparent that a solution requires development of new technologies that are probably complex and possibly require support from off-farm sectors.

Given the extent to which community-based approaches to NRM became identified with group-based extension methods, critiques of this kind led to arguments that reliance on such approaches should be reduced. For instance, Kingwell *et al.* (2008, p. 909) stated that 'there are indications that the maintained investment and emphasis on the community-based approach could be an over-investment. The community-based approach may have been more effective if more funds were directed earlier toward developing technical and economic solutions to salinity'.

Meanwhile, some researchers have been seeking to promote a broader understanding of the role of community-based approaches in Australian NRM. These researchers (e.g. Marshall 2002, 2008b; Reeve *et al.* 2002) have argued that the community-based approach is at least partly an attempt to reverse the mistrust, non-cooperation and dependency of farmers that arose from prior government-based approaches that tended to be paternalistic and unresponsive to local conditions.

Concerns have been expressed accordingly at how expectations of community-based approaches in respect of 'capacity building' became narrowed to extension efforts designed to build the human capital of individual community members, in terms of their awareness, knowledge, attitudes and skills, and at how the need to build the complementary social capital required to engender community ownership, or mutual obligation, largely became sidelined (Bellamy *et al.* 2002; Marshall 2001). [2] Lack of systematic attention to the building of such social capacities, within and between communities and governments, meant that progress in remedying the damage to these capacities from prior government-based approaches remained limited at best.

A review of Australian ICM programs in the mid-1990s, for instance, identified 'a profound lack of understanding, even a misunderstanding, about community empowerment by both government and communities' (AACM and the Centre for Water Policy Research 1995, p. 32). Price (1996, p. 5) observed that 'it is probably fair to say that the ICM process has largely been driven by government institutions … [and consequently] programs such as ICM and Landcare often have the opposite effect to that which they aspire to achieve. Many of these programs … can reinforce notions that natural resource management issues are taken care of by government programs'.

Insights from collective action theory

I have previously applied insights from the theory of collective action to help explain how devolution of governance responsibilities to community-based organisations can sometimes, given supportive conditions, increase the degree to which community members cooperate voluntarily by way of helping to discharge those responsibilities (Marshall 2004a, 2004b, 2005, 2009). This theory is relevant to community-based NRM since benefits from governance structures and their interventions are collective goods. Goods of this kind are non-exclusive in the sense that individuals contributing towards their provision are unable to exclude others from sharing the resulting benefits. Their successful provision therefore entails collective action (Olson 1965).

As explained in the next section, governance strives to help citizens provide themselves with the collective goods they seek more directly. These 'first order' collective goods in the NRM context include knowledge for defining and solving natural resource problems, and resolution of divisive intra-community conflicts. Benefits from actions that protect or enhance natural resource systems also fall into this category when they cannot be fully captured by those contributing to those actions (i.e. when they are not private goods). For instance, a farmer adopting a conservation practice that helps to enhance water quality in a river is normally unable to exclude others from sharing in the benefits. The benefits from such actions, however, do not need to constitute collective goods for a collective-action perspective to be relevant for a particular NRM problem, provided that the solution to the problem depends on governance or the provision of other collective goods like knowledge. The significance of this distinction was highlighted when Kingwell *et al.* (2003, 2008) questioned the need for community-based approaches when it was found that collective on-ground action by farmers in solving dryland salinity problems was less necessary than originally anticipated.

Olson (in Sandler 1992, p. xiii) reprised his seminal contribution to the theory of collective action (Olson 1965) by observing that 'there is an externality inherent in all collective good situations, in that each individual's provision of any amount of a collective good would confer some benefit to others'. He explained that the externality problem becomes greater the larger the group of individuals who would benefit from the good. The larger the group, the smaller the proportion that individuals capture from the benefits of their respective contributions towards provision, and thus the less motivated each will be to contribute.

Olson (1965) also observed that individuals with a greater interest in seeing a collective good provided typically contribute disproportionately to the provision effort. The expression 'free riding' was coined to describe the situation where individuals stint in their own provision efforts in the expectation that others with greater interest will contribute sufficiently that the collective good gets provided. Olson (1965, p. 62) predicted that free riding would rule in groups large enough that 'each member … is so small in relation to the total that his action will not matter much one way or the other', thus making it irrational for individual members to incur the costs of monitoring and punishing each other's free riding.

A version of this free-rider problem became interpreted by game theoreticians as an 'assurance problem' where obstacles to collective action stem from the challenges group members face in assuring one another they can be trusted to reciprocate each other's contributions (Runge 1981). Key insights into how such challenges might be overcome came from research by Axelrod (1984, p. 12) designed to test the hypothesis that: 'What makes it possible for cooperation to emerge is the fact that the players might meet again. … The future can therefore cast a shadow back on the present and thereby affect the current strategic situation.' The hypothesis was supported by the research, which demonstrated that individuals following strategies of reciprocity (i.e. who begin by contributing and then reciprocate what others do) can, if the 'shadow of the future' is strong enough, compete successfully with individuals following strategies of free riding and unconditional non-cooperation.

Nevertheless, it remained unclear how the shadow of the future could be strengthened sufficiently within large groups for a critical mass of members to become motivated to adopt reciprocity strategies. Based on an extensive review of empirical research, Ostrom (1998) concluded that individuals sharing an assurance problem engage in ongoing monitoring of each other's reputations as trustworthy reciprocators. When individuals perceive that others' adherence to a reciprocity strategy has increased, this strengthens their trust that others will reciprocate their future contributions. This strengthens the shadow of the future by raising their payoffs expected from contributing, thus strengthening their own motivations to practise reciprocity. The more those individuals come to practise reciprocity, in turn, the stronger becomes their reputations as trustworthy reciprocators.

Ostrom (1998 p. 13) concluded accordingly that 'levels of trust, reciprocity and reputations for being trustworthy are positively reinforcing'. The implication was that successful large-group provision of collective goods depends on establishing and maintaining a structural setting conducive to generating a shadow of the future strong enough that individuals become motivated to enhance their reputations as trustworthy reciprocators (Marshall 2005).

It may help to illustrate these concepts with a simple example. Let us say that the landholders in a region are experiencing water-quality problems but lack the knowledge they need to agree on the cause of their problems. This knowledge could be obtained by the landholders sharing the cost of engaging a consultant. The success of this solution depends on the landholders resisting the temptation to free ride on one another's contributions towards this cost, and reciprocating instead those contributions. The propensity of landholders to contribute rather than free ride will depend on whether they assess a critical mass of their fellow landholders to be trustworthy reciprocators. These assessments are more likely to be positive where the structural setting in their region is such that the benefits of establishing and maintaining a reputation as a trustworthy reciprocator exceed the costs of doing so.

A role for governance

One of the key elements in establishing a structural setting conducive to successful large-group collective action involves governance by a third party (North 1990).[3] Such third-party activity can bolster the shadow of the future by increasing the likelihood of free riders being identified and punished, thus preventing free riding from becoming so common that trust and reciprocity begin to unravel in a vicious cycle. Third-party governance of this kind can be organised by the group itself (e.g. by appointing members or staff to undertake this role) or it can be provided by government or some other external organisation (e.g. regional NRM organisation).

Nevertheless, governance comes at a cost. Aside from the resources expended in this activity, introducing a third party creates a new challenge of establishing and maintaining a cooperative relationship between group members and the third party on the basis of mutual trust and reciprocity. This relationship can be described as 'vertical' in the sense that the third party is situated 'above' the group members within the governance system. Hence we can distinguish between (a) vertical trust and reciprocity associated with relationships between individuals and/or organisations situated at different levels of a governance system, and (b) horizontal trust and reciprocity associated with relationships between individuals and/or organisations at the same level of a governance system. The challenge of third-party governance is to establish levels of vertical trust and reciprocity in relationships with group members that are sufficient for obtaining greater provision of the relevant first-order collective good/s than is possible with existing stocks of horizontal trust and reciprocity alone.

The more that vertical trust and reciprocity is lacking between group members and the third party, and, consequently, the less that members cooperate voluntarily with its monitoring and enforcement efforts, the greater the need for costly coercion and the risk of successful collective action becoming unaffordable (Marshall 2002, 2005, 2008a). Although focused on governments as third parties, the following remarks from Ostrom (2009, p. 8) are valid for third parties generally:

> *The problem of collective action does not disappear once a government makes a policy to deal with an externality. Governmental policies need to rely to a great extent on willing cooperation from citizens. When citizens approve of a government policy, think they should comply, and this view is complemented by a sense that the government policy is effective and fairly enforced, the costs of that enforcement are much lower than when citizens try to evade the policy.*

The focus in the literature on the vertical dimension of the challenge faced in large-group collective action has been largely on the importance of gaining and maintaining group members' trust in the third party. The presumption tends to be that increasing this trust will increase the degree to which group members cooperate voluntarily with third-party decisions. In their review of factors affecting rural landholders' adoption of conservation practices, Pannell *et al.* (2006) concluded that a key determinant of landholders' adoption of conservation practices is their trust in the researchers who developed the practices and in the advisers who promote it to them, and presumed that adoption would tend to be influenced positively by such trust.

In previous research (Marshall 2004a, 2004b, 2008a, 2009), however, I have explained that the relationship between farmers' vertical trust in, and their vertical voluntary cooperation with, third-party governance is more complex than normally assumed. The relationship depends on the kind of strategy that farmers are following in their interactions with the third party.

There will be no relationship between farmers' vertical trust and their vertical cooperation when farmers are following strategies that are independent of their levels of vertical trust; for example, when they are opposed to cooperating regardless of their trust in the third party (Marshall 2008a, 2009). The relationship will be negative when they are following strategies of free riding in their interactions with the relevant governance structures. Farmers following free-rider strategies with a community-based organisation supporting them to adopt relevant conservation practices cooperate less with that organisation the more they trust it as a source of reliable support (Marshall 2008a, 2009).

Free-riding behaviour takes various forms. A situation that is commonly observed is where farmers accept financial support for establishing conservation works on the understanding that they will maintain the works, but then stint in this maintenance in the expectation that further support will become available to rehabilitate or replace the works (Gibson *et al.* 2005). Barr and Cary (1992) found that expectations in Australia of farmers maintaining government-funded conservation works at their own expense have been normally misguided. Farmers can also free ride on support in various other ways, including: not adopting practices as agreed (e.g. in return for financial support) unless strong enforcement occurs; curtailing existing plans for on-ground NRM activity in the expectation that financial support for that activity will later become available; relaxing one's efforts to learn about conservation problems and practices in the expectation that advice will be provided through extension activities; and reducing their participation in local groups working on the kinds of problems for which the support is provided. This last example illustrates the risk of existing horizontal cooperation between farmers becoming weakened when third-party governance is applied in ways (e.g. paternalistically) that motivate the farmers to free ride on the support it provides.

When farmers follow strategies of vertical reciprocity in their interactions with the relevant governance structures, in contrast, the relationship between their vertical trust and their vertical cooperation will be positive. Farmers following strategies of vertical reciprocity with a community-based organisation supporting them to adopt conservation practices would coop-erate more with that organisation the more they trusted it as a source of reliable support (Marshall 2008a, 2009). This cooperation could take various forms, including: accepting financial support only for adoption that would not have otherwise occurred; maintaining con-servation works as agreed; adopting practices as agreed in the absence of enforcement, or beyond the minimum levels that were agreed; exerting peer pressure on other farmers to adopt the practices; establishing local farmer groups to work more effectively on shared problems; and acquiring information about NRM problems and practices in excess of that gained through extension activities.

In recent publications from this research (Marshall 2008a, 2009), it was argued that caution is required in assuming that Australian farmers nowadays generally follow reciprocity strate-gies in their interactions with governance structures, community-based or otherwise, respon-sible for supporting them to adopt conservation practices. Hence:

> *With the Australian history of paternalistic natural resources governance in respect of farmers, and the considerable antagonism this has sometimes caused, it seems reasonable to assume that few farmers were following reciprocity strategies in their dealings with government before the mid-1980s. This pattern may have begun to shift around this time as a consequence of the introduction of Landcare, integrated catchment management, and other community-based programs. With such programs, Australian governments turned towards supporting the self-reliance of farmers and their communities in addressing natural resource issues, while expecting farmers and their communities to reciprocate this support by voluntarily contributing some resources towards resolving these issues (e.g. investing in adoption of recommended on-farm conservation practices).*
>
> *Nevertheless, the focus of such programs until the introduction of the regional delivery model was on community levels no higher, relative to the scale of regions delineated under this model, than what is now called 'subregional'. To the extent that such programs have effected a transition in farmers' dealings with higher-level NRM bodies towards strategies of reciprocity, therefore, we would expect this transition to be more evident in their*

dealings with subregional groups than in their dealings with regional groups (Marshall 2009, p. 1515).

The Samaritan's dilemma

The challenge of engaging farmers in relationships of vertical reciprocity with organisations responsible for promoting their self-reliance, instead of creating perverse incentives for farmers to become more dependent, tends to be more readily acknowledged in the literature on 'third-world' rural development (Gibson *et al.* 2005; Gronemeyer 1992; Korten 1983; Schumacher 1964; Uphoff *et al.* 1998) than it has been in related 'first-world' public policy discourse. The relevance of this challenge for third-world environmental management was recognised by Pretty and Ward (2001, p. 220) who wrote of the need 'to ensure the transition is made from dependence [of individuals and their groups] to interdependence'. In a recent discussion of this challenge in the third-world context, Ellerman (2007, p. 563, original emphasis) wrote that:

The assumed goal is transformation towards autonomous development on the part of the doers, with the doers helping themselves. The problem is how can the helpers supply help that actually furthers rather than overrides or undercuts the goal of the doers helping themselves? This is actually a paradox. If the helpers are supplying help that is important to the doers, then how can the doers really be helping themselves? Autonomy cannot be externally supplied. And if the doers are becoming autonomous, then what is the role of the external helpers? This paradox of supplying help to self-help ... is the fundamental conundrum *of all helping relationships.*

Nevertheless, the relevance of this paradox for first-world contexts has not gone unnoticed by scholars. Buchanan's (1977, pp. 169–185) discussion of the 'Samaritan's Dilemma' was a seminal contribution in this respect, and one which allows us to elaborate the collective-action perspective on the role of community-based NRM that was considered earlier. His game theory model of the 'active' version of this dilemma involves two players, the 'Samaritan' (or 'helper') and the 'recipient' (or 'doer'). The relevance of this model for natural resources governance has been noted recently by Gibson *et al.* (2005) and Bruns (2008). In this model, the helper has the choice of providing, or not providing, help to the doer. The doer can choose between reciprocating, or not reciprocating, the help by helping herself. Given the configuration of payoffs to the two players in this model, the helper will choose to help regardless of the choice she expects the doer to make. In contrast, the doer's choice depends on the helper's choice. If the helper chooses to not help, the doer maximises her payoff by choosing to help herself. If the doer knows the helper will always choose to help, however, the former will always choose against helping herself.

Nevertheless, the helper would prefer the outcome where she helps and the doer reciprocates by helping herself. Buchanan (1977) reasoned that the helper could achieve this outcome by recognising that her choice affects the doer's choice, and then behave 'strategically' by pretending that her preferred response to the doer not helping herself is to withhold help. Such strategic behaviour will be costly to the helper, however, who denies her true preferences whenever she denies help. Holding to this behaviour therefore requires from the helper what Buchanan called 'strategic courage' (Buchanan 1977, p. 173) and parents nowadays may refer to as 'tough love'.

Buchanan was pessimistic about the prospects for strategic courage in modern societies. He reasoned that the increasing wealth of such societies was undermining the motivation for such courage, given that it 'may be a markedly inferior economic good … As incomes have increased, and as the stock of wealth has grown, men have increasingly found themselves able to take the soft options' (Buchanan 1977, p. 173).

The Samaritan's dilemma, community-based NRM, and farmer adoption of conservation practices

What does the prior discussion add to our understanding of the role of community-based NRM in strengthening farmers' adoption of conservation practices? To begin with, two key lessons can be drawn from Buchanan's analysis. Firstly, the tendency for farmers to reciprocate help (e.g. extension services or financial incentives) from NRM governance structures (e.g. government agencies or community-based organisations) can be expected to strengthen with their perceptions that those structures possess the strategic courage needed to practise vertical reciprocity with them. The more that farmers expect future help to be withheld if they do not reciprocate present help (e.g. by adopting conservation practices promoted to them through such help), for example, the more likely are they to reciprocate the help rather than free ride on it and thereby become more dependent.

Farmers can be expected to reciprocate help only when they perceive it as helpful. It is not uncommon in NRM for actions intended to help farmers to be perceived by them as unhelpful. Accordingly, 'help' is defined here in terms of farmers' perceptions, rather than according to the helper's intentions. An advantage often claimed for community-based NRM, compared with centralised NRM governance, is that community-based structures have better access to local knowledge and are better able to adapt their actions to local context. To the extent that this is true, actions by community-based structures intended to help farmers are more likely to be perceived as help by those farmers.

The second key lesson is that NRM governance structures face considerable challenges in convincing farmers that they possess strategic courage of this kind. Farmers have a number of reasons to be sceptical of the ability of governments, or other governance structures, to summon and sustain such courage.

One reason derives from farmers' historical experience of governments acting paternalistically towards them (i.e. and therefore not recognising possibilities for reciprocal community self-help where they existed) on the presumption that community members are generally incapable of contributing usefully to NRM efforts. Another reason is that governments often face strong political pressures to 'take action' on NRM problems, and therefore can be reluctant to desist from action if farmers' reciprocation fails to eventuate. A third reason is that governments not uncommonly evaluate their agencies and other structures they fund on their demonstrated ability to expend funds within the intended time frame. Hence, governance structures can be discouraged from exercising strategic courage because withholding help may lead to under-expenditure of their budgets, and thus to reduced budgets in subsequent years.

A fourth reason for farmers' scepticism regarding the ability of NRM governance structures to summon and sustain strategic courage may stem from their awareness of the self-interest of those involved in the business of providing help. Ellerman (2007, p. 565) noted how:

> The helping professions do depend on neediness, disability, incapacity, and helplessness to make their living so they are in the paradoxical position of working to eliminate their own jobs – at least insofar as they actually try to help people help themselves.

A fifth reason stems from the fact that governance structures are often in competition with one another. This can undermine each structure's strategic courage by leading to expectations that the benefits of their own courage will be negated by other structures compensating for any help they withhold. This reason relates to the well-known phenomena of 'turf protection' and 'empire building'. A sixth reason derives from the high costs often encountered in monitoring individual farmers' actions such that a legitimate basis exists for judging that reciprocal self-help is lacking, and consequently that future help should be withheld. Farmers' capacities in the short term to reciprocate any help they receive can vary greatly with fluctuations in their seasonal and market circumstances. Farmers following reciprocity strategies who face severe seasonal and market circumstances can be expected to postpone their reciprocation until their circumstances improve. Hence, judgements of whether farmer self-help is lacking can require a longer-term perspective in order to be regarded as legitimate. The final reason to be mentioned here relates to the high transaction (including political) costs that can arise in enforcing decisions to withhold future help from farmers whose reciprocal self-help is judged lacking. These costs can be high given the continuing effectiveness of farmers in the political arena.

Towards a deeper appreciation of what 'community' means for farmer adoption of conservation practices

What scope exists then for a community-based approach to (1) increase strategic courage within the NRM governance system sufficiently that farmers become motivated to follow reciprocity strategies in their interactions with the system, or (2) reduce the strategic courage needed to provide this sufficient level of motivation?

Appropriate devolution of NRM governance responsibilities to community-based organisations may strengthen strategic courage in fulfilling those responsibilities to the extent that community members leading and staffing those organisations are likely more motivated to see the particular NRM problems faced by their community solved than would be the case in larger-scale organisations with a more diffuse interest in seeing those problems solved. A number of case studies in Australia have observed this effect. Based on her research into catchment, Landcare and other environmental stewardship groups under NHT1, Carr (2002, p. 123) observed how:

> Once stewardship groups are successful in securing financial resources and begin implementing projects, they tend to pore over the allocation and disbursal of funds at a level of detail much more precise than that of normal government accounting systems. … It could be argued that group members have a higher stake in ensuring that their money is well spent than the anonymous government employee has in protecting government coffers.

Marshall (2002, 2004a, 2004b, 2005) found similar dynamics at work in his case study of the devolution of NRM governance responsibilities to Murray Irrigation Ltd, a local company co-owned by the irrigators to which it supplies water. An officer of the Murray-Darling Basin Commission involved in the devolution exercise observed, 'It's interesting that when you get things down to an arrangement, communities tend to be tougher on themselves than they'll let government be with them' (Marshall 2004b, p. 161).

We might also expect devolution of NRM responsibilities to community-based organisations to reduce the level of strategic courage needed within the governance system to ensure fulfilment of those responsibilities. A number of reasons may be advanced for this expectation. The first reason is that decomposing a given responsibility (e.g. deciding how government funds should be invested in particular NRM activities) so it can be devolved to multiple lower-level groups (e.g. from a state government agency to a series of regional community-based organisations) can break down a large-group problem of collective action into a series of smaller-group problems more conducive to group members cooperating voluntarily with one another in helping to fulfil that responsibility. The more that a responsibility is discharged by group members self-reliantly in this way, the less is the need for the governance system to provide help in order for the responsibility to be discharged satisfactorily. It follows that less vertical reciprocity is required from group members (i.e. farmers) in their interactions with the governance system, and the need for that system to exercise strategic courage is reduced accordingly. Hence, a given stock of strategic courage is more likely to be adequate for discharging a particular responsibility.

This reasoning is consistent with a key lesson drawn by Axelrod (1984) from his research into the role of reciprocity in successful collective action. This lesson, 'Enlarge the shadow of the future' (Axelrod 1984, p. 126), was based on his finding that 'mutual cooperation can be stable if the future is sufficiently important relative to the present. This is because the players can each use an implicit threat of retaliation against the other's defection – if the interaction will last long enough to make the threat effective'.

One of the basic ways that Axelrod recommended for enlarging the shadow of the future was 'to make interactions more frequent. In such a case the next interaction occurs sooner, and hence the next move looms larger than it otherwise would' (Axelrod 1984, p. 129). He reasoned that interactions between specific individuals can be made more frequent by making interactions between those individuals more concentrated, and remarked on how having multi-levelled organisation of people promotes voluntary cooperation by making interactions between specific individuals more concentrated.

A second reason why the need for strategic courage may be reduced by devolving NRM governance responsibilities to community-based organisations is that this devolution may strengthen community ownership of those responsibilities. Where this strengthening is sufficient, the 'payoffs of the game' may shift from those of a Samaritan's Dilemma towards payoffs more aligned with farmers reciprocating help provided to them. This reason is consistent with another of the key lessons that Axelrod (1984, p. 133) drew from his research into collective action problems, namely, 'change the payoffs'.

Dynamics of this kind were observed in the aforementioned case study of devolution of NRM governance responsibilities to Murray Irrigation Ltd. An officer with the NSW Government involved in the process judged that the devolution of responsibilities to this community-based organisation had been 'very important in getting real change on farms. You would not be able to get it out of Government' (Marshall 2004b, p. 161).

A third reason for why devolving responsibilities may reduce the need to establish strategic courage within a NRM governance system is that this devolution brings the system closer to the world of farmers, and thus can make it easier for them to monitor the degree of strategic courage that does exist in that system. This effect may be particularly important where the objective is to 'turn around' an NRM governance system that has established a reputation among farmers for low strategic courage. The easier it is for farmers to observe that a governance system is acting with increased strategic courage, the less is the need for this strengthening to be 'acted out' in order for farmers to recognise that strengthening has indeed occurred.

This reasoning is consistent with another of the key lessons that Axelrod (1984, p. 139) drew from his research: 'the scope of sustainable cooperation can be expanded by any improvements in the players' ability to recognise each other from the past, and to be confident about prior actions that have actually been taken' (Axelrod 1984, p. 140).

Some quantitative evidence

The reasoning in the previous section suggests that involving the community in NRM governance may mean considerably more for farmer adoption of conservation practices than embracing a more participatory approach to developing the individual capacities (or human capital) of farmers in terms of the awareness, knowledge, attitudes and skills that they need for this governance to succeed. It indicates also that significant additional scope exists for a community-based approach to strengthen the collective capacities of the NRM governance system (particularly in the form of vertical trust and reciprocity between farmers and the governance structures seeking to strengthen their self-reliance in NRM) to solve problems in this domain. This is not to suggest that individuals' awareness, knowledge, attitudes and skills are irrelevant for these collective capacities. Rather, the suggestion is that these collective capacities will remain underdeveloped while the task of strengthening vertical social capital in the governance system remains neglected.

Up to this point in the chapter, however, this reasoning is yet to be corroborated by quantitative evidence. This shortfall is addressed below, where relevant quantitative evidence from two research projects is discussed. The focus in this discussion is on examining whether farmers in the cases studied were following strategies of vertical reciprocity in their interactions with the community-based NRM organisations helping them through extension activities and financial incentives to adopt locally-relevant conservation practices.

Two projects covering four cases of community-based NRM

The first of the research projects was referred to in the previous section. It involved a case study of devolution of NRM responsibilities to Murray Irrigation Ltd, a company co-owned by the irrigators to whom it supplies water. Details of the case study context are available from Marshall (2001, 2002, 2004a, 2004b). The second of the projects involved case studies in three NRM regions defined under the regional delivery model. These regions (and subregions) are larger and more populous than the four irrigation districts in the first project. See Marshall (2008a) for further details of the three cases studied in the second project.

The first of the three regions was the South West Catchments. Within this region, the focus was on the Blackwood Basin. The regional and subregional NRM bodies were the South West Catchments Council and the Blackwood Basin Group, respectively. Another of the regions was the Fitzroy Basin Region in Queensland. Here, the focus was on the Central Highlands subregion. The regional and subregional NRM bodies in this case were the Fitzroy Basin Association and the Central Highlands Regional Resources Use Planning Cooperative, respectively. The remaining region was the Mallee Region in Victoria. To maintain comparability across the three cases, the focus in this region was on dryland farming districts since agricultural activity in the other two focal subregions was predominantly dryland-based. The regional NRM body in this case was the Mallee Catchment Management Authority, which decided against establishing subregional arrangements with similar status as in the other two cases.

Research method

The quantitative research method followed in each project was similar, and involved:

(1)   surveying random samples of farmers

Table 8.1.   Details of the surveys.

Project and case	Timing of survey	No. of farm businesses responding	Approx. no. of farm businesses in population
Project 1: Central-Murray Irrigation Districts (NSW)[a]	Jul–Aug 1999	235	1610
Project 2: Blackwood Basin (Western Australia)[b]	Sept 2006–Feb 2007	333	1950
Project 2: Central Highlands (Queensland)[b]	Sept 2006–Feb 2007	170	890
Project 2: Dryland areas of Mallee NRM Region (Victoria)[b]	Sept 2006–Feb 2007	318	862

[a] (Source: Marshall 2004a.)
[b] (Source: Marshall 2008a.)

(2)   using the survey data to measure (a) each respondent's intended change in adoption of conservation practices promoted to them by their community-based NRM organisation, (b) each respondent's level of trust in their community-based organisation as a source of NRM assistance, and (c) values of other 'control' variables

(3)   testing statistically whether the relationship between (a) and (b) across each sample of respondents is positive as would be expected if farmers were predominantly following reciprocity strategies in interacting with their community-based organisation.

Step (3) in each case involved application of multiple regression methods. In testing any relationship between a dependent variable (e.g. farmers' intended adoption of conservation practices) and a particular explanatory variable (e.g. farmers' trust in their community-based organisation as a source of NRM assistance), these methods control for the influence on the dependent variable of other explanatory variables expected to be relevant (e.g. farmers' current financial viability, education levels, and perceptions of whether adoption would serve their interests).

An explanatory variable in respect of farmers' current financial viability was included in each regression model to control for the influence on farmers' adoption plans of their current financial circumstances. Otherwise, the degree to which farmers in a region facing severe circumstances (e.g. prolonged drought) at the time of their survey were actually following reciprocity strategies over the longer term may have been underestimated. The risk that the reciprocity of farmers facing current negative circumstances may be underestimated was addressed also by measuring their adoption-change intentions over a period long enough (10 years) that reasonable opportunities to reciprocate the support provided to them should have arisen. Details of the surveys undertaken for the four case studies are in Table 8.1.

Findings from the central-Murray case (first project)

The survey item used in the first project to gauge farmers' trust in their community-based organisation involved them rating their level of agreement with a statement that Murray Irrigation Limited was committed to supporting implementation of the 'land and water management plans' (LWMPs) that had been developed through community-based processes in each of the four irrigation districts under its jurisdiction. Multiple regression analysis identified a statistically significant positive relationship between the variable constructed from this survey item (called *trust in Murray Irrigation*) and farmers' intended levels of adoption of conservation practices relevant to their situations for which aggregate adoption targets (i.e. for farmers as a group) had been set in the LWMPs (Marshall 2004a). We can conclude, therefore, that farmers in this setting were predominantly following reciprocity strategies with their community-based NRM organisation.

Findings from three case studies under the regional delivery model (second project)

A combination of five survey items was used in the second project to gauge farmers' trust that their relevant community-based regional NRM organisations were committed to supporting them in adopting the kinds of conservation practices prioritised in their region's NRM strategy and investment plans. The explanatory variable constructed from these items was called *trust in regional body*. An equivalent set of five items was used to measure farmers' trust that their relevant subregional organisation was similarly committed (except in the Mallee dryland case where no such organisation existed). The explanatory variable derived for the two relevant cases from this set of items was called *trust in subregional body*.

In each of the three cases, multiple regression models were estimated for each of the conservation practices that the relevant community-based organisation was promoting to farmers as a matter of priority. With 22 such practices identified across the three cases, this number of models was estimated. The dependent variable for each model related to each farmer's expected change in adoption (measured in hectares) of a particular conservation practice over the subsequent 10 years (Marshall 2008a, 2009).

Fifteen models were estimated in the two cases where *trust in subregional body* was relevant. Nine of these 15 models (60%) identified a statistically significant positive relationship between this explanatory variable and farmer's expected change in adoption of the relevant practice. None of the remaining six models identified a statistically-significant negative relationship. Hence, we can conclude that farmers in these two cases were (a) predominantly following reciprocity strategies with their subregional community-based NRM organisation for more than half of the practices promoted to them, and (b) predominantly not following free-riding strategies in respect of the remaining practices promoted to them.

In contrast, six of the full array of 22 models (27%) identified a statistically-significant relationship between *trust in regional body* and farmers' expected change in adoption of the relevant practice. Aside from the one Mallee dryland model in this set, the five other models found the relationships to be negative. In the Blackwood Basin and Central Highlands cases, therefore, it seems that farmers' trust in their regional NRM body is, when it is exerting influence, predominantly influencing farmers' adoption plans through free-rider dynamics.

Interestingly, four of the five models from the Blackwood Basin and Central Highlands cases that found *trust in regional body* to have a significant negative relationship with farmers' expected adoption changes also found *trust in subregional body* to have a significant positive relationship with farmers' expected adoption changes. The implication seems to be that subregional bodies in these cases have been more successful than their counterparts at the regional level in managing the Samaritan's Dilemma – and thus in leading their relationships with farmers away from those fostering dependency towards those that foster self-reliance.

The Mallee model in this set found that the relevant regional body (Mallee CMA) had managed predominantly to foster relationships of reciprocity with farmers in respect of the relevant conservation practice. This CMA's greater apparent success compared with the two others in fostering relationships of reciprocity with farmers might be explained by the absence of a subregional group in this case. This absence may have led farmers in this case to interact more directly with the CMA than otherwise would have occurred, and thus allowed farmers greater opportunity to recognise from its behaviour that (at least in respect of this practice) it possesses the strategic courage needed to motivate reciprocity from them.

Nevertheless, the Mallee CMA seems to have established reciprocity dynamics with farmers for a lower proportion of priority conservation practices than was established by the subregional bodies in the other two cases. A possible explanation is that community-based NRM at the regional level is more remote from the world of farmers than is a community-based approach at the subregional level. This may disadvantage a regional body, compared with a subregional body, in two ways:

(1)   in establishing a reputation for strategic courage[4]
(2)   in gaining from farmers the ownership of its initiatives needed to 'change the payoffs of the game' sufficiently for them to become motivated to help implement those initiatives on the basis of reciprocity.

Discussion of findings

The quantitative evidence presented in this section from two different research projects, and from the four different cases that these projects covered, indicates consistently that reciprocity was the predominant strategy followed by farmers with their community-based NRM organisations in these cases for a substantial share of the relevant conservation practices. Nevertheless, we lack 'control treatments' (i.e. government-based programs providing similar NRM support in similar contexts) with which to compare these cases. Hence, it is not possible to determine conclusively that any occurrence of farmers following reciprocity strategies in interacting with their community-based organisations is a consequence of introducing a community-based approach.

However, it seems likely given the historical experiences of Australian farmers with governmental NRM interventions that farmers would predominantly not be following reciprocity strategies if they were now dealing with government-based programs (Marshall 2008a, 2009). The Natural Resources Commission (2008, p. 13) observed accordingly that:

> While there may be some administrative efficiency gains to be made from centralising control [of NRM governance from CMAs], they are likely to be overwhelmed by the loss of community support and inability of central agencies to motivate the long-term behaviour change that CMAs are beginning to drive.

We can conclude with reasonable confidence from the quantitative evidence considered above that the reciprocity found to be followed by farmers when interacting with their community-based NRM bodies is a consequence of introducing a community-based approach. The second of the two research projects (which focused on the regional delivery model) generated some quantitative evidence also for the proposition that the ability of a community-based approach to engage farmers in relationships of reciprocity with their community-based organisation tends to increase the more that responsibilities under the approach are devolved towards the local level. However, further evidence is needed for this proposition to be evaluated with reasonable confidence.

Conclusions

The reasoning and evidence presented in this chapter combine to indicate that the conventional understanding of the role of community-based NRM in Australian agricultural landscapes – as a group-based approach to rural extension concerned with farmers' relevant awareness, knowledge, attitudes and skills – underestimates the potential of this approach. This is not to deny that farmers' self-reliance in NRM can be enhanced when community-based approaches help them more effectively to develop these elements of their human capital. However, it is clear from the preceding sections that attempts to help develop farmers' self-reliance in NRM can be counterproductive unless community-based approaches succeed in developing the complementary elements of 'vertical' social capital needed in the NRM governance system (i.e. vertical trust and reciprocity between farmers and NRM governance structures) to motivate farmers to reciprocate the help provided to them. Without both these elements of social capital, there is a real risk that farmers will predominantly free ride on the help they receive rather than reciprocate it, so that their dependence on external help is reinforced or deepened.

The role of social capital in fostering farmers' adoption of conservation practices has not gone unnoticed in the literature on community-based NRM in Australia. However, discussions in this area have tended to focus on one particular element of social capital, i.e. farmers' trust in the organisations promoting conservation practices to them through extension activities and financial incentives. This literature has largely presumed that farmers' adoption of conservation practices invariably tends to increase with their trust in the organisation promoting the practices to them. In contrast, it is evident from the logic and evidence presented in this chapter that the influence of this vertical trust of farmers on their decisions to adopt conservation practices depends on whether they are motivated to practise reciprocity – at least over the longer term – with organisations supporting them in adopting those practices. In other words, it is not enough that farmers trust their regional NRM bodies to support them in addressing the NRM problems they face. Farmers' self-reliance will be strengthened under such arrangements only to the extent that this trust exists and farmers have become motivated to reciprocate the support they receive.

Successful community-based NRM requires leaders and policy makers at all levels to deepen their understandings of how this approach strengthens farmers' self-reliance in adopting conservation practices. This perspective complements the conventional view of community-based NRM based on the ideas of rural extension by bringing in insights from developments in the theory of collective action.

These insights reveal the promise of a community-based approach in strengthening the motivation of farmers to reciprocate the support given to them towards adopting relevant conservation practices, and thus in truly enhancing their self-reliance in adopting these practices. However, this promise will be realised only to the extent that leaders and policy makers at all levels:

- acknowledge the Samaritan's Dilemma normally faced in helping farmers become more self-reliant, and accordingly
- attend to developing community-based arrangements with the 'strategic courage' needed to motivate most farmers to reciprocate the support they receive towards adopting conservation practices.

The responsibility in a community-based organisation for establishing the vertical social capital needed to motivate farmers to reciprocate the organisation's support to them cannot be left to a 'community' program which has limited influence over how the rest of the organisation behaves. Farmers' trust in the organisation depends on their experiences with it across all its activities, as do their perceptions of its strategic courage. A whole-of-organisation strategy for establishing vertical social capital with farmers needs to be followed for a community-based organisation to realise its promise in helping farmers to more effectively help themselves.

Meanwhile, we need to keep the promise of community-based NRM in perspective. Strengthening farmers' motivations to adopt conservation practices will not be enough when the practices on offer are contrary to their interests and are thus not in the vicinity of being 'adoptable'. Hence, realising the promise of community-based NRM often relies on complementary efforts to develop more adoptable practices (Pannell *et al.* 2006). Here too, external support for such efforts (e.g. government investment in research) will likely encounter the Samaritan's Dilemma. Appropriate involvement of community-based organisations in these efforts can be a way of addressing the risk of these efforts undermining farmer motivations to undertake complementary efforts of their own.

Acknowledgements

Preparation of this chapter was supported by a grant from the Commonwealth Environmental Research Facilities program for the project 'Improving economic accountability when using decentralised, collaborative approaches to environmental decisions'. Neither this program nor the Australian Government necessarily shares the views expressed in the chapter. Feedback from Tim Cummins, Warren Musgrave, Elinor Ostrom, Ian Reeve and Alistair Watson was greatly appreciated in preparing this chapter. The author is responsible for remaining deficiencies.

Endnotes

1   Regional NRM organisations in New South Wales are called Catchment Management Authorities (CMAs).
2   A typical regional NRM strategy addresses a number of NRM themes (e.g. land, water, biodiversity, and community) of which one is normally called the 'community' theme or similar. 'Community capacity-building' activities undertaken within such a theme tend to be dominated by extension activities focused on farmers' human capital in terms of their awareness, knowledge, attitudes and skills.
3   As distinct from 'first-party' monitoring, enforcement and other governance activities undertaken directly by group members.
4   Both because (a) the regional body's greater remoteness from farmers in a particular community may lead to a more diffuse interest in seeing their specific NRM problems solved, and (b) the regional body's greater remoteness increases the difficulty that farmers face in monitoring its behaviour closely enough to gauge its strategic courage.

References

AACM and the Centre for Water Policy Research (1995) 'Enhancing the effectiveness of catchment management planning'. Final report. Department of Primary Industries and Energy: Adelaide.

Agriculture Fisheries and Forestry Australia (1999) 'Managing natural resources in rural Australia for a sustainable future: a discussion paper for developing a national policy'. Agriculture, Fisheries and Forestry Australia: Canberra.

Axelrod R (1984) *The Evolution of Cooperation.* Basic Books: New York.

Barr NF and Cary JW (1992) *Greening a Brown Land: The Australian Search for Sustainable Land Use.* Macmillan: Melbourne.

Bellamy JA, Ross H, Ewing SA and Meppem T (2002) *Integrated Catchment Management: Learning from the Australian Experience for the Murray-Darling Basin.* CSIRO Sustainable Ecosystems: Brisbane.

Bruns BR (2008) 'Aiding adaptive co-management in irrigation'. Paper presented at the 12th Biennial Global Conference of the International Association for the Study of the Commons, July 14–18, Cheltenham, England.

Buchanan JM (1977) *Freedom in Constitutional Contract: Perspectives of a Political Economist.* Texas A&M University Press: College Station.

Carr A (2002) *Grass Roots and Green Tape: Principles and Practices of Environmental Stewardship.* Federation Press: Sydney.

Cary J, Webb T and Barr N (2002) *Understanding Landholders' Capacity to Change to Sustainable Practices: Insights about Practice Adoption and Social Capacity for Change.* Bureau of Rural Sciences: Canberra.

Chambers R (1983) *Rural Development: Putting the Last First.* Longman: London.

Cunningham GM (1988) Total catchment management. *Journal of Soil Conservation* **44**(1), 42–45.

Curtis A (1998) Agency-community partnership in Landcare: lessons for state-sponsored citizen resource management. *Environmental Management* **22**(4), 563–574.

Curtis A, Lockwood M and MacKay J (2001) Exploring landholder willingness and capacity to manage dryland salinity in the Goulburn Broken catchment. *Australian Journal of Environmental Management* **8**(2), 79–90.

Curtis A, Lucas D, Nurse M and Skeen M (2008) *Achieving NRM Outcomes through Voluntary Action: Lessons from Landcare.* Department of Sustainability and Environment: Melbourne.

Curtis AL (1997) Landcare, stewardship and biodiversity conservation. In *Frontiers in Ecology: Building the Links.* (Eds NI Klomp and I Lunt) pp. 143–153. Elsevier: Oxford.

Curtis AL and De Lacy T (1996) Landcare in Australia: does it make a difference? *Journal of Environmental Management* **46**(2), 119–147.

Ellerman D (2007) Helping self-help: the fundamental conundrum of development assistance. *Journal of Socio-Economics* **36**(4), 561–577.

Esman MJ and Uphoff NT (1984) *Local Organisations: Intermediaries in Rural Development.* Cornell University Press: New York.

Garrett P, MP (2008) Interview transcript at launch of the Caring for our Country: Outcomes 2008–2013 Statement.

Gibson CC, Ostrom E, Andersson K and Shivakumar S (2005) *The Samaritan's Dilemma: The Political Economy of Development Aid.* Oxford University Press: Oxford.

Gronemeyer M (1992) Helping. In *The Development Dictionary: A Guide to Knowledge As Power.* (Ed. S Wolfgang) pp. 51–69. Zed Books: London.

Hollick M (1992) Why won't they do it? Problems of implementing catchment management at the farm level. In *Proceedings of the 5th Australian Soil Conservation Conference.* (Eds GJ Hamilton, KM Howes and R Attwater) pp. 51–55. Western Australian Department of Agriculture: Perth.

House of Representatives Standing Committee on Environment, Recreation and the Arts (1989) 'House of Representatives inquiry into land degradation: the effectiveness of land degradation policies and programs'. Australian Government: Canberra.

Industry Commission (1998) 'A full repairing lease: inquiry into ecologically sustainable land management'. Industry Commission: Canberra.

Kingwell R, Hajkowicz S, Young J, Patton D, Trapnell L, Edward A, Krause M and Bathgate A (Eds) (2003) 'Economic evaluation of salinity options in cropping regions of Australia'.

Report to the Grains Research and Development Corporation and the National Dryland Salinity Program: Canberra.

Kingwell R, John M and Robertson M (2008) A review of a community-based approach to combating land degradation: dryland salinity management in Australia. *Environment, Development and Sustainability* **10**(6), 899–912.

Korten DC (1983) Social development: putting people first. In *Bureaucracy and the Poor: Closing the Gap*. (Eds D Korten and F Alfonso) pp. 201–221. Kumarian Press: West Hartford.

Lockie S and Vanclay F (Eds) (1997) *Critical Landcare*. Centre for Rural Social Research, Charles Sturt University: Wagga Wagga.

Marsh SP and Pannell DJ (2000) Agricultural extension policy in Australia: the good, the bad and the misguided. *Australian Journal of Agricultural and Resource Economics* **44**(4), 605–627.

Marshall GR (2001) Crafting cooperation in the commons: an economic analysis of prospects for collaborative environmental governance. PhD thesis. School of Economics, University of New England: Armidale.

Marshall GR (2002) Institutionalising cost sharing for catchment management: lessons from land and water management planning in Australia. *Water, Science and Technology* **45**(11), 101–111.

Marshall GR (2004a) Farmers cooperating in the commons? A study of collective action in salinity management. *Ecological Economics* **51**(3/4), 271–286.

Marshall GR (2004b) From words to deeds: enforcing farmers' conservation cost-sharing commitments. *Journal of Rural Studies* **20**(2), 157–167.

Marshall GR (2005) *Economics for Collaborative Environmental Management: Renegotiating the Commons*. Earthscan: London.

Marshall GR (2008a) *Community-based, Regional Delivery of Natural Resource Management: Building System-wide Capacities to Motivate Voluntary Farmer Adoption of Conservation Practices*. Rural Industries Research and Development Corporation: Canberra.

Marshall GR (2008b) Nesting, subsidiarity, and community-based environmental governance beyond the local level. *International Journal of the Commons* **2**(1), 75–97.

Marshall GR (2009) Polycentricity, reciprocity, and farmer adoption of conservation practices under community-based governance. *Ecological Economics* **68**(5), 1507–1520.

Mues C, Chapman L and Van Hilst R (1998) 'Survey of Landcare and land management practices: 1992–93'. Australian Bureau of Agricultural and Resource Economics: Canberra.

Natural Resource Management Ministerial Council (2002) 'National natural resource management capacity building framework'. NRM Ministerial Council: Canberra.

Natural Resource Management Ministerial Council (2006) 'Framework for future NRM programmes'. Natural Resource Management Ministerial Council: Canberra.

Natural Resources Commission (2008) 'Progress report on effective implementation of catchment action plans'. Natural Resources Commission: Sydney.

North DC (1990) *Institutions, Institutional Change and Economic Performance*. Cambridge University Press: Cambridge.

Olson M (1965) *The Logic of Collective Action*. Harvard University Press: Cambridge.

Ostrom E (1998) A behavioral approach to the rational choice theory of collective action. *American Political Science Review* **92**(1), 1–22.

Ostrom E (2009) 'Nested externalities and polycentric institutions: must we wait for global solutions to climate change before taking actions at other scales?' Working Paper 09–5, Workshop in Political Theory and Policy Analysis. Indiana University: Bloomington.

Pannell DJ, Marshall GR, Barr N, Curtis A, Vanclay F and Wilkinson R (2006) Understanding and promoting adoption of conservation practices by rural landholders. *Australian Journal of Experimental Agriculture* **46**(11), 1407–1424.

Pretty J and Ward H (2001) Social capital and the environment. *World Development* **29**(2), 209–227.

Price R (1996) 'Integrated catchment management: national questions to be resolved'. Paper presented at the Integrated Catchment Management Forum: Effective ICM in an Urbanising Catchment, Radisson Rum Corp Resort.

Reeve I, Marshall GR and Musgrave W (2002) 'Resource governance and integrated catchment management'. Issues Paper No. 2 for Murray-Darling Basin Commission project MP2004. Institute for Rural Futures, University of New England: Armidale.

Runge CF (1981) Common property externalities: isolation, assurance, and resource depletion in a traditional grazing context. *American Journal of Agricultural Economics* **63**(4), 595–605.

Sandler T (1992) *Collective Action: Theory and Applications*. University of Michigan Press: Ann Arbor.

Schumacher EF (1964) A humanistic guide to foreign aid. In *Development and Society*. (Eds D Novack and R Lekachman) pp. 364–374. St Martin's Press: New York.

Senate Standing Committee on Rural and Regional Affairs and Transport (2010). Natural resource management and conservation challenges. Canberra: Australian Government.

Steering Committee (2000) 'Steering Committee report to Australian governments on the public response to "Managing natural resources in Australia for a sustainable future: a discussion paper for developing a national policy"'. Agriculture, Fisheries and Forestry, Australia: Canberra.

Uphoff N, Esman MJ and Krishna A (1998) *Reasons for Success: Learning from Instructive Experiences in Rural Development*. Vistaar Publications: New Delhi.

Vanclay F (1992) The social context of farmers' adoption of environmentally sound farming practices. In *Agriculture, Environment and Society: Contemporary Issues for Australia*. (Eds G Lawrence, F Vanclay and B Furze) pp. 94–121. Macmillan: Melbourne.

Wallington TJ, Lawrence G and Loechel B (2008) Reflections on the legitimacy of regional environmental governance: lessons from Australia's experiment in natural resource management. *Journal of Environmental Policy and Planning* **10**(1), 1–30.

9

I hope you are feeling uncomfortable now: role conflict and the natural resources extension officer

Neil Barr

Summary

There is an implicit social contract between farmers and their advisers. A key clause of this contract is that advisers will provide information and advice that is of benefit to farmers. Historically, when the role of government-employed extension in Australia was primarily to promote the adoption of practices that were intended to increase the profitability of farms, it was relatively easy for advisers to comply with this contract. However, the shift in emphasis of government-funded extension away from productivity and towards conservation outcomes has made it more difficult. Advisers now can find themselves in a position where their employer or funder demands that they promote activities that they and farmers know are not compatible with the farmer's goals. The result can be a loss of credibility and trust, with important implications for the way that extension agents are viewed by farmers. They are unlikely to be allowed to participate in farmers' decision making in a supporting role, or even to help farmers evaluate information. Rather, they are limited to providing information that other more trusted individuals assist the farmer in evaluating this information and deciding how to respond to it. Thus, government extension has evolved into a less influential activity than it once was. To counter this, funders need to develop more sophisticated strategies that acknowledge and integrate conflicting farmer and government objectives. The alternative is for their extension service to lose relevance to the farming community.

Introduction

One of the strangest meetings I have ever attended was with a group of pasture and grazing industry extension staff seeking to understand why they were expected to advise farmers to do something that made no sense. Most farmers would agree that advice to sow perennial pasture during a prolonged drought is generally bad advice. These extension staff felt they were being asked to give such advice to justify their employment.

No-one set out to create this situation. One could say it was an emergent outcome of this particular funding program and its associated organisational structure (Bella 1997). To understand, it required following the money trail. Although the extension staff were paid by their employer, a State Department of Agriculture, the money came from a catchment body together with a contract to promote the sowing of perennial pasture. The catchment body was interested in the salinity and erosion control benefits of increased perennial pasture coverage in the catchment. This increased coverage was to be achieved using a combination of persuasion, demonstration and incentive payments. The catchment authority was but one intermediary in the money trail. The money came to the authority from the Federal Government under a national conservation program. Like any national program, the ultimate funding source had accountability requirements that were passed down the chain. The national funding body was under pressure from an audit office report questioning the effectiveness of the program. By the time this pressure reached the extension officers, it was expressed as an expectation to achieve a fixed area of newly sown perennial pasture for each year of the project. Accountability reporting was an annual occurrence.

Unfortunately, severe drought arrived in the midst of this project's delivery timetable. The target audience understood that any money invested in sowing would be wasted and would increase the financial pressure on the farm business during a difficult season. Extension staff who advocated the sowing of pasture to managers of drought-stricken farms would damage their local credibility. The pressures of annual reporting placed the extension officers in an impossible position of having to defend their failure to achieve annual sowing targets to the catchment body. The unfortunate message they inferred was that they should be concerned only with the objectives of the funding agent, rather than take account of the situation of the farm community in drought. But one could not conclude that government was not interested in the welfare of farmers, farm families and farm businesses during the drought. Other federal funding programs were providing financial support to the same farm community during the drought.

One can allocate responsibility for this situation at a number of steps along the funding chain. Perhaps pasture adoption was not the most effective investment available to the catchment body to help it achieve its objectives. Perhaps extension investment was not the most effective strategy for achieving the pasture adoption objective. Perhaps the objective of extension investment needed to be justified on the basis of building the capacity to sow pasture when suitable seasonal conditions returned. These questions are discussed in the recent Australian literature on targeting natural resource investment (Pannell 2008). This chapter leaves these questions for others, and instead examines the transformation of the role of the extension officer as government priorities have shifted from agricultural production to conservation. One of the consequences of this shift has been the transformation of the social contract that underlies the farm advisory relationship.

This chapter starts with a short exploration of the nature of the perennial pasture innovation. It then recounts some important research into the advisory relationship. Some of this research is now more than 20 years old. Despite the age of this research, it offers some useful insights into the nature of the social contract between farmer and adviser. This social contract sets the bounds of reasonable performance expectations of any investment in advisory services.

Conservation and complexity

Early in my career I found myself working in a program promoting dryland lucerne for watertable control on the riverine plains of northern Victoria. The simple message of the program seemed to be 'plant lucerne; it will solve your watertable problems!' Fortunately, there were other advantages to growing lucerne. It was potentially more profitable than the usual pastures of the area. However, at that time only a few farmers grew significant areas of dryland lucerne (Ransom and Barr 1993). It took an explanation by the local lucerne system specialist for the problem of complexity to become apparent. What looked like a simple act of sowing an alternative pasture was potentially the first step in a transformation of the farming system.

Farmers sowing lucerne did not have a guarantee they would produce a successful crop. The chance of failure was greater than for most other pasture species. One way to minimise the financial risk of establishing lucerne, and to make up for time a paddock may be out of production was to sow lucerne with a faster growing crop such as safflower. Farmers following this strategy had to learn to grow new crops which were more compatible with lucerne (Barker 1992).

Lucerne requires rotational grazing management. The majority of farms were managed with a regime of set stocking. Wool-producing farms typically ran three flocks: ewes, weaners and wethers. Some ran an additional flock of maiden ewes. Under the four-paddock rotation system recommended for dryland lucerne, a farm keeping the same flock structure would need 12 or 16 paddocks. For farms previously 'set-stocked' this implied additional expensive fencing and more dams and water reticulation to provide watering points in each paddock. Fencing at this intensity was likely to impede the easy management of cropping activity on the farm. Only larger farms built up over a number of farm amalgamations had an existing paddock structure suitable for rotational grazing. Farmers sowing dryland lucerne needed large farms, or needed to rethink their flock structure.

Lucerne pasture is more productive than normal pasture, but wool producers did not make money merely by growing more pasture. More sheep would be required to utilise the extra pasture (Ransom 1992). The increased flock size required extra working capital, more work in sheep handling and an increased workload of rotational grazing. Higher sheep densities in paddocks may mean a greater need for control of intestinal parasites and increased use of veterinary chemicals or greater attention to rotational grazing systems to minimise parasite infestation (Coffey 1992).

One means of maximising the benefits of lucerne is to abandon lambing in autumn in favour of spring lambing and convert to prime lamb production. This meant a need to further rearrange the farm timetable. Shearing would probably be moved to after the harvest season and before sowing. The risk of grass seed contamination would increase. Grazing rotation strategies to minimise this risk were needed. To maximise the benefits of prime lamb production, the farmer would need to develop new marketing skills and develop relationships with export abattoirs.

Most farms in the region were mixed farms. These changes to the grazing side of the enterprise needed to be integrated into the cropping enterprise. Growing lucerne can mean major changes in crop management. How does the farmer combine the new grazing rotation with the crop rotation side of the business? Whereas an annual pasture may have been grazed for a couple of years before cropping, there are good reasons to maintain a lucerne paddock for its full eight-year life after successful establishment. Consequently, the farmer may have to crop paddocks elsewhere on the farm for a longer period before putting them back into pasture. Forestalling the depletion of soil nitrogen inevitably meant introducing grain legumes and oil seeds into a rotation system that was predominantly based on wheat and pasture. This required improved cropping skills, marketing skills and probably investment in cropping machinery and on-farm storage for each crop.

Clearly, the 'simple' decision to sow dryland lucerne was not all that simple. It implied complex changes to the farm management system and necessarily entailed greater risk in the conversion period. There were many issues to think through, and only some of the changes were capable of being tested using discrete farmer experiments. In short, this was an adoption decision that required deep thought and assessment of conflicting information. There may be significant and undesirable consequences for the farm business if it did not work out.

Adoption of farm innovations with these characteristics is never fast. It may take a generation between first experiments and widespread adoption across a farming region, as has been the experience with conservation cropping. During this period there will be phases of disadoption as new problems with the system are discovered, and re-adoption of these practices as those problems are solved (Pannell *et al.* 2006). The role of the extension agent in promoting such technologies is complicated and subtle.

Making decisions as a social engagement

In the 1980s a graduate student of agricultural extension at Melbourne University undertook a longitudinal anthropological study of the process of farmer decision-making. Tom Phillips conducted multiple interviews during a year in the life of a number of dairy farmers who were thinking about big changes to their farming business (Phillips 1985). This may have been the first rural 'sense-making' research in Australia.

Tom represented the decision-making environment with three concentric circles (see Figure 9.1). At the centre of the circle was the decision-making farmer. In the inner circle around the farmer was family and those who would share the consequences of any decision. In the middle circle were trusted associates. In the outer circle were socially distant subject experts. The search for information to support a decision is shown by the arrows. In this hypothetical example, the decision maker starts with a trip to an information expert in the outer circle, then confers with family intimates, makes another trip to an information expert, seeks help to evaluate the information in the middle circle, then returns to the inner intimates circle to consider the decision.

In a series of interviews with each farmer, Tom recorded the process of decision making. As well as documenting the decision journey, he asked farmers to identify and classify their face-to-face information sources, and then he mapped the information-seeking behaviour of the

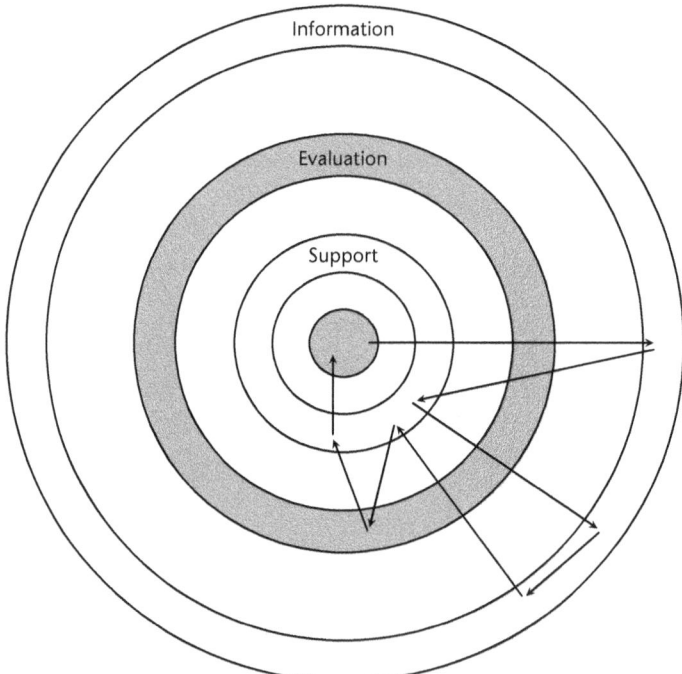

Figure 9.1.   Tom Phillips' conceptual model of decision making.

(From Phillips 1985.)

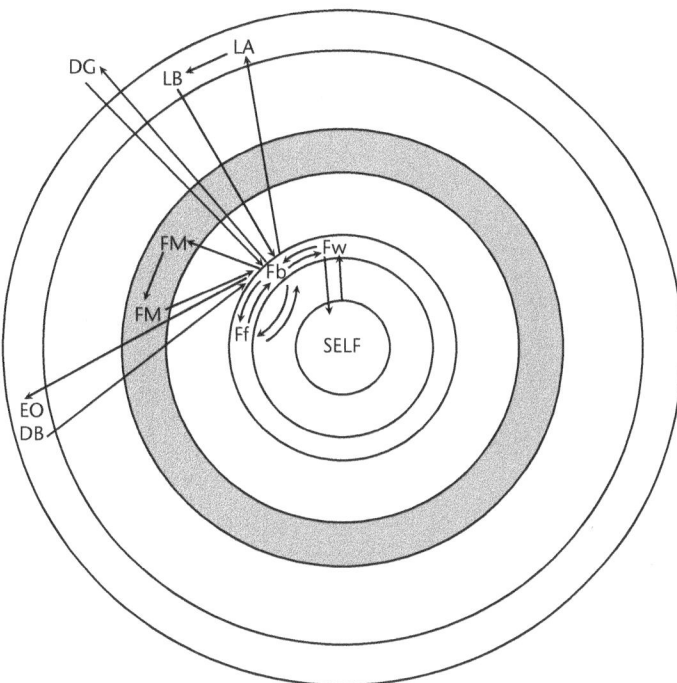

Figure 9.2.   Petal diagram representing a decision about fencing and subdivision.

(From Phillips 1985.)

decision maker between sources and confidents located in each of these circles. The average farmer undertook 30 learning projects in the year. Time available for learning was constrained. This meant time was allocated according to the significance of the learning project to the core values and aspirations of the farmer. For small and inconsequential decisions, the decision maker often undertook a quick assessment of the available information and made a decision. An example of making a decision about fencing and paddock subdivision is represented by Figure 9.2. In this situation, information providers are approached (LA, LB, DG, EODB), and one person (FM) is used to evaluate the information. The final decision making is done with three family members (Fl, Fb, Fw).

For decisions that carried significant risks for the farmer, the process of decision making was generally long and extensive. Farmers invariably described these decisions as stressful. Research in cognitive psychology reveals that such decision making is uncomfortable, often less than rational, and embedded within the social support network of the decision maker. Most people dislike uncertainty, and when faced with a decision with high uncertainty and complexity, are likely to apply simple decision rules (heuristics) rather than find an optimal solution (Tversky and Kahneman 1974). Commonly used heuristics include the recognition heuristic (choosing the familiar) and the imitation heuristic (copy others). Humans also fear losses far more than they value potential gains (Kahneman and Tversky 1984). Faced with a complex decision with potential for significant loss, we humans have an in-built tendency to stick with social norms and the *status quo* (Samuelson and Zeckhauser 1988).

These human traits mean that the process of consequential decision making is more than just an information-seeking task. It is a sense-making task and a search for emotional support (Janis and Mann 1977). A decision about buying a block of land and building a new dairy shed

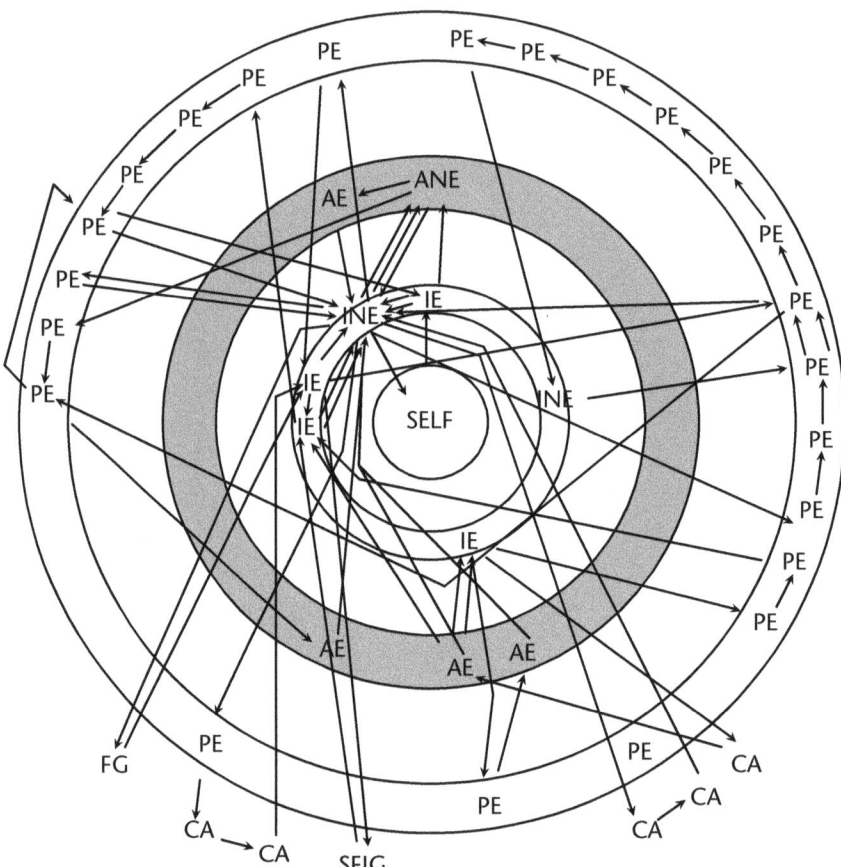

Figure 9.3.   Petal diagram portraying a decision to build a new dairy shed.

(From Phillips 1985.)

is shown below (Figure 9.3). There were many excursions to the outer circle for information, but also significant trips to people in the inner and middle circles. I will refrain from describing the process for obvious reasons.

Decision makers undertake different tasks in each of the circles. In the outer circle, he or she is seeking information from many sources. There is more and more information available to farmers. How does one make sense of this information? The evaluation of the information is performed in the middle circle. Here it is evaluated, integrated and sense is made of the information. This task is shared with a small number of contacts. Finally, decisions are made in consultation with the inner circle. These are the people who will share in the outcomes of the decision. Here the decision maker is looking for emotional as well as intellectual support to evaluate the options against personal and family goals. The issues will not only be 'will this work?', but also 'how will these people react to my decision, and will they support me if I make this decision?'

The more difficult the decision, the more the stages of information seeking and evaluation will be intertwined. The decision maker will engage in a series of sorties, engaging and re-engaging the personal support network and less intimate sources of information. The major decision will be preceded by a series of decision points along the way. At each of these points the decision maker will validate the decision with close or intimate contacts, with more time spent with closest contacts when the main options are evaluated (Phillips 1985).

For the extension (or practice-change) agent seeking to promote the adoption of complex system innovations such as lucerne, there is clearly greatest opportunity if one is included in the farmer's middle circle of contacts where information is assessed and integrated. But an invitation into the inner circle must be earned by the adviser.

Another researcher provided some insights into who is included in this middle circle and why they are included. Anderson (1979, 1981) studied the characteristics of extension workers which made them more or less acceptable and credible to farmers. It was important for the advisor to be technically capable of taking a farm system view of any technology and to deduce practical options and assessments. The credibility of extension agents depended upon more than just technical systems competence. Just as crucial was that the farmer believed that the agent accepted the primacy of the farmers' goals. Those who were seen as aiming to help the farmer work towards his or her objectives were trusted. Those with different agendas were not. It is easy to see that any extension agent seen as working towards goals irrelevant to or even in conflict with the farmer's goals will at best be relegated to the outer circle of information sources, to be just one of many potentially competing information sources (Vanclay 2004).

Common goals

Many farm operators have common objectives – of being full-time farmers, of providing a reasonable standard of living for their families and passing on a viable farm business to the next generation. A farm that meets these objectives needs large enough scale to generate a cash surplus to provide an acceptable standard of living for the family (potentially supplemented with off-farm income from a partner), allow the building of financial reserves to smooth income fluctuations caused by commodity prices and climate fluctuations, and generate the capacity to invest in expansion and productivity improvement to keep pace with the declining terms of trade.

In my current work, I ask gatherings of farmers across Victoria about the scale of farm needed to meet these objectives. The belief is that one needs a farm capable of generating an annual gross income of approximately $400 000 during average seasonal conditions. The number varies according to the industry and the group. While these assessments are subjective, data from the ABARE farm survey supports this simple rule of thumb.[1]

Figure 9.4 shows a time series of mean farm cash surplus and non-farm income for Victorian broadacre farms with a gross farm income of between $100 000 and $200 000. The mean farm cash surplus only rises above $50 000 in a couple of years. When non-farm income is included and if we assume the farm contributes to only one household budget, the average total annual income into the farm household over the past five years has been $61 000. This is a little below the median Australian household annual gross income of $68 800 in 2008 (ABS 2009). It is difficult to see how farms of this scale can enable a farmer to achieve the multiple objectives of providing an acceptable standard of living and allow for investment to mitigate income fluctuation and investment for future productivity increases. The long-term future for farms of this scale will include serial dependence upon government assistance during periods of low rainfall or low commodity prices. The long-term prognosis is that these farms will fall further behind as the terms of trade decline.

Figure 9.5 shows the same time series for farms with a farm gross income of between $200 000 and $400 000. The anecdotal benchmark for financially sustainable production lies at the upper end of the farm scale within this group. Mean farm cash surplus varied between

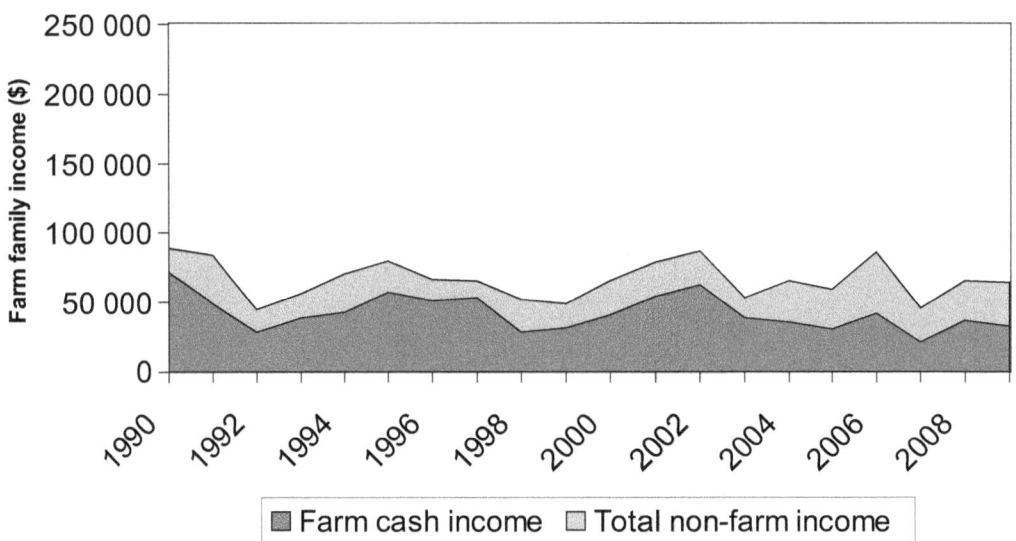

Figure 9.4. Farm cash income and total non-farm income for Victorian broadacre farms with gross income between $100 000 and $200 000 for the period 1990–2008.

(Source: ABARE Agsurf database, http://www.abare.gov.au/interactive/agsurf/.)

$50 000 and $120 000, and was supplemented by a modest non-farm income. Over the most recent five years of the series, the average total annual income from cash surplus and non-farm income was $86 000. If we make the same assumptions about single household farms, we can equate this income to the 70th percentile of Australian household incomes. This naïve analysis suggests there is capacity to sacrifice some household income, maintain an average standard of living and make some investment in the farm and to smooth income volatility. The anecdotal benchmark of farmer groups suggests this is only really feasible for the very largest farms in this group.

Farm scale in Australia is strongly skewed. There are many small farms and far fewer larger farms. Figure 9.6 shows the distribution of farm financial scale in Victoria in 2006. Between 20 and 25% of farm establishments are on a scale that surpasses the anecdotal benchmark for long-term financial sustainability. In other words, three-quarters of the farms in Victoria fall below this benchmark. For many of these farms, we can generally assume that any decision to invest in farm technology raises trade-offs between the needs of the farm and need for household expenditure. Many of the smallest of these farms are managed by farmers in a form of retirement or by operators dependent upon off-farm employment. The owners of these farm businesses have little capacity to take on any business risk. Most of these farms are located in areas where land values are high, making farm expansion through land purchase an unlikely strategy (Barr and Karunaratne 2002). Because of age or off-farm work commitments, few are able to consider changes to their farm system that significantly increase labour requirements from either the owner operator or from hired labour. The only pasture innovations likely to be compatible with these farmers objectives are those with low cost, low risk and low complexity (Barr 1996; Barr and Wilkinson 2006). Promoting any other pasture system will probably conflict with many of these farmers' objectives.

For farms a little larger in scale (between $150 000 and $300 000 EVAO), there is less dependence upon off-farm income and, in average seasons, should provide an acceptable

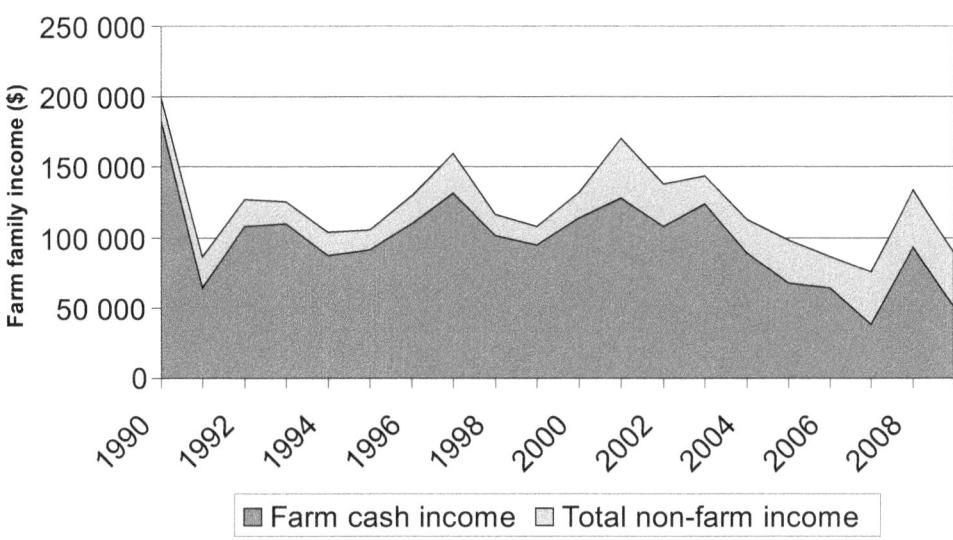

Figure 9.5.   Farm cash income and total non-farm income for Victorian broadacre farms with gross income between $200 000 and $400 000 for the period 1990–2008.

(Source: ABARE Agsurf database, http://www.abare.gov.au/interactive/agsurf/.)

family income. However, these medium-sized farms have limited capacity to invest in farm development and expansion for a sustainable future. For many operators the objective is to

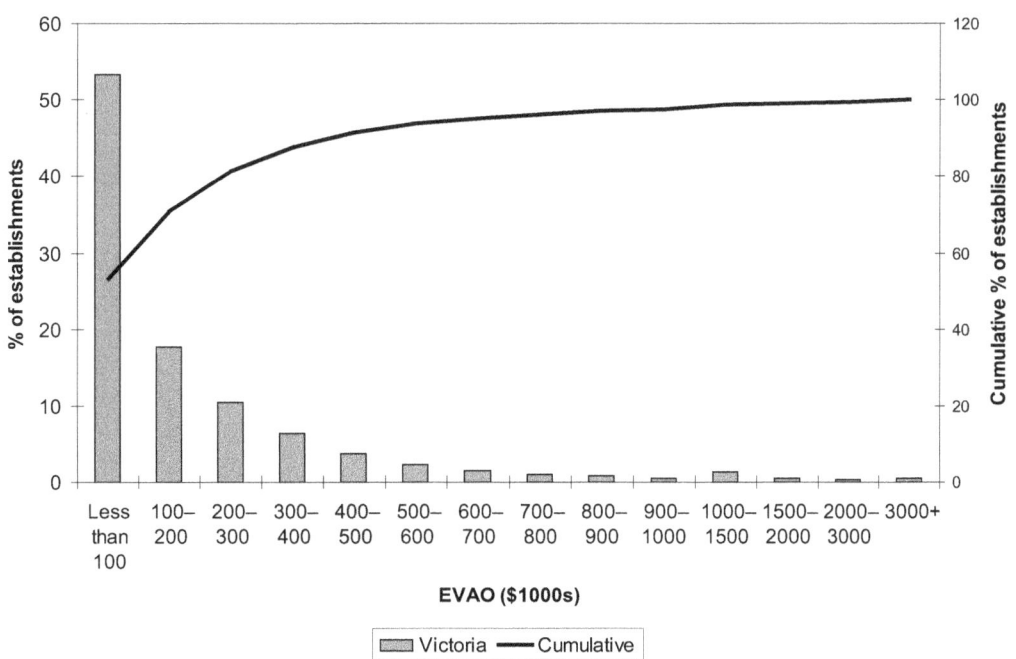

Figure 9.6.   Farm scale distribution and cumulative distribution for Victoria in 2006.

(Data source: ABS, derived from customised data tables from the 2006 Australian Agricultural Census.)

maintain viability for as long as possible, at least until retirement. All technology will be viewed through this lens. Technology with financial risk will be unattractive for many who judge they have minimal capacity to absorb losses from failed changes to their farm systems.

Serving two masters?

Extension providers fall along a spectrum between the counselling form of extension and 'practice-change' extension. In the counselling school of farm advice, acceptance of the goal framework of the farmer is seen as instrumental as a means of achieving an extension objective of improved decision making, and as a desirable ethical objective. During the 1970s and 1980s, this approach gained considerable influence in the extension profession with extension being portrayed as a 'helping profession' by its practitioners. The goal of the extension agent was to assist the client to achieve their goals through improved decision making. From this perspective, it was argued that measures of adoption were not useful indicators of extension success. The area of pasture sown was less important than whether farmers made sensible decisions as to whether to sow pasture or not. This philosophy of extension continues today amongst the Rural Financial Counselling service and private consultants.

This counselling form of advisory support delivered by government advisors was gradually withdrawn through the 1980s and 1990s as government objectives shifted from rural development towards the containment of agricultural externalities. Extension officers could no longer justify their activity through 'improved decision making' but through promises of tangible adoption outcomes that delivered wider public benefits. For a short period, the increasing divide between farmer objectives and government conservation advisors was addressed by projects that attempted to 'change farmer attitudes' (Barr 1994). Thankfully this approach is now discredited, though the cause of discredit has been its lack of effectiveness in changing adoption rates rather than any ethical objection (Cary 1994).

The phrase 'extension' is now being gradually replaced by phrases such as 'practice-change'. This is at least overt in making it clear that the government extension officer's objectives may be different to those of the farmer. The implicit admission of the potential for diverging agendas between government and advisory services inevitably makes the social contract position of the advisory officer much more complex. Funding accountability requires the government advisory officer (or the government contracted consultant) to be accountable to the funder for contracted outcomes, such as the adoption of perennial pasture systems. But to be effective in achieving this outcome, the officer needs to maintain a credible standing with the farmers working through this adoption process. And the key to that is to respect the objectives of the farmer. Role conflict is now likely to be an inevitable experience for the government extension agent.

Endnote

1   The anecdotal benchmark can be confirmed using ABARE Farm Survey data. One assumes that a farm operator has three basic objectives: to be a full-time farmer, to earn enough to provide a living standard commensurate with that of the rest of the community and to pass on a viable business to the next generation. Based upon the past decade, the farm income will need to keep pace with a 1% annual decline in the terms of trade and a 2% per annum real increase in disposable income in the wider community. Based upon linear regression relationships between cash surplus, capital and farm gross receipts, one can calculate the average scale of farm that is capable of meeting all these criteria. In 2008 this was somewhere approaching $500 000 gross receipts. A spouse working off-farm can lower this threshold to approximately $400 000.

References

ABS (2009) 'Household income and income distribution, Australia, 2007–08'. Australian Bureau of Statistics: Canberra.

Anderson AM (1979) *How Advisors Advise: Agricultural Extension As a Social Process*. Hawkesbury Agricultural College: Richmond.

Anderson AM (1981) *Farmers' Expectations and Use of Agricultural Extension Services*. Hawkesbury Agricultural College: Richmond.

Barker J (1992) *Lucerne Intercropping*. Centre for Land Protection Research: Bendigo.

Barr N (1994) Landcare from inside-out and outside-in. *The Australian Farm Manager* **5**(1): 2–10.

Barr N (1996) Conventional and low input pasture improvement: a review of recent market research. *New Zealand Journal of Agricultural Research* **39**(4), 559–568.

Barr N and Karunaratne K (2002) *Victoria's Small Farms*. Natural Resources and Environment: Bendigo.

Barr N and Wilkinson R (2006) Social persistence of plant-based management of dryland salinity. *Australian Journal of Experimental Agriculture* **45**(11), 1495–1501.

Bella DA (1997) Organized complexity in human affairs: the tobacco industry. *Journal of Business Ethics* **16**(10), 977–999.

Cary JW (1994) Modelling beliefs related to environmental behaviour. Unpublished PhD thesis. University of Melbourne: Melbourne.

Coffey M (1992) Farmers' perceptions of upgrading pasture to realise the potential of their grazing land. 4th year B.Ag.Sci. project. La Trobe University: Bundoora.

Janis M and Mann L (1977) Decision making: a psychological analysis of conflict, choice and commitment. Macmillan: New York.

Kahneman D and Tversky A (1984) Choices, values and frames. *American Psychologist* **39**(4), 341–350.

Pannell DJ (2008) Public benefits, private benefits, and policy mechanism choice for land-use change for environmental benefits. *Land Economics* **84**(2), 225–240.

Pannell DJ, Marshall GR, Barr N, Curtis A, Vanclay F and Wilkinson R (2006) Understanding and promoting adoption of conservation practices by rural landholders. *Australian Journal of Experimental Agriculture* **46**(11), 1407–1424.

Phillips TI (1985) The development of methodologies for the determination and facilitation of learning for dairy farmers. Master's thesis. School of Agriculture and Forestry, University of Melbourne: Melbourne.

Ransom K (1992) *Dryland Lucerne Grazing Systems: Development, Demonstration and Computer Simulation of Grazing Systems to Utilise Mixed Pastures of Sub Clover, Annual Grasses and Dryland Lucerne*. Department of Agriculture: Bendigo.

Ransom K and Barr N (1993) *The Adoption of Dryland Lucerne in North-Central Victoria*. Department of Agriculture: Bendigo.

Samuelson W and Zeckhauser R (1988) Status-quo bias in decision-making. *Journal of Risk and Uncertainty* **1**(1), 7–59.

Tversky A and Kahneman D (1974) Judgement under uncertainty: heuristics and biases. *Science* **185**(4157), 1124–1131.

Vanclay F (2004) Social principles for agricultural extension to assist in the promotion of natural resource management. *Australian Journal of Experimental Agriculture* **44**(3), 213–222.

10

Women in agriculture

Cathy McGowan

Summary

This chapter highlights the work of women in agriculture and on the farm, and the implications for extension and practice change. It encourages the reader to move beyond the stereotype of 'the man on the land', 'the head of the household', and the farmer as 'he', to pay attention to all the work done in the context of the family farm business. It reviews the relevant literature around women's work in agriculture, discusses the barriers to their participation, and provides practical strategies for effectively engaging all those involved in family farm businesses. It argues that when innovation and new ways of working are required, involving women in the design, participation and evaluation of programs will pay dividends.

Introduction

'*Do you have a gender element in your extension program?*', the farming woman asked the older, experienced head of extension in the Department of Primary Industries. He responded smartly and unperturbed: '*We don't do gender, we do people!*' (Quote, July 2007).

The purpose of this chapter is to explore the implications of this question and the ramifications of the answer for practice change and extension in Australian agriculture. The main contention is that, traditionally, the work of women within the family farm business has not been understood, counted or acknowledged, and consequently has fallen outside the scope of mainstream research, extension and change management activities.

In 2000, the Women in Dairy Leadership Project won an Australasia Pacific Extension Network (APEN) award for excellence in extension. This project worked with women in the dairy industry, encouraging and supporting them to take leadership roles in the industry. There were three key lessons from that project which have relevance for practice change. Gender does matter. Successful participation programs are those which encourage and support women's involvement and which address the barriers to their participation. Effective change management programs acknowledge diversity at all levels of program design, implementation, management and review. This chapter expounds and explains these three simple yet important messages. It is the author's belief that involving all members of the family farm business team in relevant activities is a simple, low-cost method to bring about long-term change and enhance uptake of new practices.

Gender matters

In a book on practice change and extension, why is it necessary to have a chapter about women? Why isn't it enough to think of people as 'people' and treat them all the same? Is there really a problem in saying '*We don't do gender, we do people*'? The simple answer is that if we want change, then there is a problem, because women and men are different.

While it sounds obvious to say that men are different from women, or that young people differ from old people, these are truisms that are often ignored in the design of programs. People see the world differently and act from different frameworks. Taking these differences into account is an important element in gaining participation, in winning hearts and minds, and in creating environments for long-term sustainable change. To be effective in the process of enabling change, it is important to recognise and manage these differences, and to address the hidden and institutionalised barriers described later in this chapter.

Helen Cixous (cited by Moi 2002) argues that there are two ways of being in the world, the 'masculine' and the 'feminine', although neither is necessarily a matter of biology. She calls the masculine the 'economy of the proper', and proposes that it tends to be preoccupied with property, appropriation and propriety, with establishing and defending borders, and processing and controlling what lies within them. In contrast, she calls the feminine the 'economy of the gift', and argues that it likes to cross boundaries, to give and receive, and to live from within, open and attentive to what is unspoken and often unspeakable. By and large, I belong to this later economy, and in my experience the relatively low levels of feminine qualities in the public life of agriculture is a loss to those involved in the sector.

There are many pragmatic reasons for taking a proactive approach to considering gender in extension activities. For example, if the whole team is involved in learning, greater gains can be made than if only one team member is involved (Kilpatrick and Bell 2000). This is because if all team members are involved from the beginning, they are better able to make shared decisions, rather than one member trying to convince other members of the benefits of a new approach. During the process of implementation, team members also provide support to one another to assist with problem solving and developing the confidence to move forward.

With 98% of Australia's farm businesses run by farm families (Garnaut and Lim-Applegate 1998; Vanclay 2003), women form a significant part of the agricultural workforce. From agricultural statistics we know that some 32% of Australia's farm workforce is female. This means that there are approximately 70 000 women in Australia who are farmers or farm managers. In economic terms, their work contributes around $4 billion annually to the economy (Elix and Lambert 1998b).

There is also the matter of fairness. '*Nothing about us, without us*' was the catch-cry of the 4th International Conference of Women in Agriculture held in South Africa in 2007. Women want to be involved. They want to be included in the processes of decision making and information gathering that will have impacts on their livelihoods, their workloads and the future of their families and communities.

The invisible farmer

In 1992, a farmer and educator, Julie Williams, working from the Warragul Adult Education Centre in Gippsland in Victoria with funding from the (then) Commonwealth Department of Primary Industries and Energy, laid the foundations for 'counting' women in agriculture. Her report, *The Invisible Farmer: A Summary Report on Australian Farm Women* (Williams 1992), was the first major report to document current information on the contribution of Australian farm women to agriculture. It was a landmark report which provided a theoretical and practical basis to the work of emerging organisations such as Australian Women in Agriculture (AWiA). It also augmented the establishment of women's units within most state departments of agriculture which began to occur in the 1980s.

Williams (1992) noted that farming had traditionally been a male domain, with women being seen as the homemakers without a significant or visible role in agriculture. The reasons for this were:

- women's unpaid work was not recognised as legitimate work
- women were excluded from agricultural education opportunities up until the 1970s (and in various ways subsequently as well)
- for women who took on leadership roles, there was isolation, lack of support and negative peer group pressure
- reinforcement of traditional stereotypes by the rural press.

Despite these barriers, in 1986 there were reportedly 78 589 women farmers and farm managers, and a further 13 178 women working as farm hands and assistants (Williams 1992). Williams questioned the accuracy of these statistics and the processes used to document women's work by statisticians and in surveys. Her concerns were further supported by the research undertaken by New Zealand politician and academic, Marilyn Waring, whose book, *Counting for Nothing: What Men Value and what Women are Worth*, had been published in 1988. Waring's finding was that the way the national accounts were calculated deliberately ignored women's work, and consequently most of women's work would never be acknowledged for the important role it played in the economy.

Research into inequality between men and women in rural Australia undertaken by Ken Dempsey (1992), a sociologist from La Trobe University, and published as *A Man's Town,* demonstrated further the existence of a gender divide. The conclusion from these various reports was that there was a serious problem in our understanding of the impact of gender and an important first step would be to collect accurate and reliable information to be 'disaggregated by gender'. This call was finally answered with the publication in 1998 by the Rural Industries Research and Development Corporation (RIRDC) of the *Missed Opportunities* reports. Volume 1 defined the roles played by women in agriculture in the agricultural sector, and Volume 2 was an economic analysis of the estimated current and potential contribution of women in agriculture (Elix and Lambert 1998a, 1998b). 'In 1995–96, the National Accounts report that the market value of farm input was $14.5 billion. When you include the value of household work, volunteer and community work and off-farm wage income earned by people on farms, the real farm income was just over $28 billion. Women contribute 48% of this real farm income' (Elix and Lambert 1998b, p. 53).

Being invisible wasn't only a matter of poor data being collected. Women themselves had contributed to this invisibility. It is partly an issue around words, meanings and contexts. The replies given by many farm women when asked the question, 'What do you do?' – such as '*Oh, I don't work, I just help out on the farm*', or '*I don't work, I am only a housewife*' – helped reinforce a stereotype of women who do little serious farm work. In the early days of the movement of women in agriculture, the need to address how women described their work was a high priority. How best should we describe the complicated work carried out by women on farms? The term, 'farmer's wife', was not a popular option. Other descriptors such as women farmers, women who farm, farming partners, or members of family farm businesses, were seen by many as being little better. The discussion continues.

Hidden in our language and thinking are also assumptions about what a 'real' farmer is. Traditionally, the 'farmer' has been the person whose name is on the land title, the man, the older man, and the head of the house. The assumptions behind these terms are that there is only one person responsible for 'farming', and he is male. The farmer by definition is a man. He is the man on the land (Silvasti 2003a, 2003b; Vanclay *et al.* 2007). Even when it has been understood that a team of people is engaged in the family farm business, the notion has still persisted that it is the men who are farmers, who are the landholders, who go to field days, who turn up at farmers' meetings, and stand for election to be farmer representatives in the agricultural and farmer organisations. A consequence of the invisibility of women is that the main target for agricultural research and extension has been 'the male farmer' and his needs. Thus, a significant opportunity for engaging with all farmers has been missed.

It is worth noting that there were a number of significant exceptions to this experience of invisibility, particularly single women farmers and career women. Many of these women were either unmarried or widowed. They have played important roles both at the farm level, in agricultural politics, and as researchers and extension agents within the various departments of agriculture.

What is the work of farm women?

Women's journey into the world of farming is typically different to that of men. Women usually come to farming through marriage. Firstly, it's a love relationship. It may then transform into a family relationship and perhaps a business relationship. The woman is likely to have had a previous life in another community. She may have completed formal training in ancillary areas such as nursing, teaching, secretarial work, hairdressing, medicine, law or journalism. On the other hand, most men come to farming as a result of their birth position. Traditionally, it is a son who inherits or partly inherits the family land. His career typically is more practical and hands-on; he has learnt the craft from his father, uncle or grandfather. His informal training is likely to have been supplemented by completion of some accredited courses, and his focus will be more on the production aspects of the farming business.

These different career paths into farming have resulted in a high level of gender specialisation in the workforce and in the family unit. The question of women's work and men's work has now been well researched and there is strong data which can be used to inform the design and implementation of all types of programs.

In 1995, Jane Gooday led the way with her survey of women in Australian broadacre and dairy farms. She noted that women have a diversity of roles and responsibilities within the farm business. This research was supplemented by Jayne Garnaut and colleagues with their analysis of ABARE statistics – in 1998 *People in Farming* and in 1999 *Farmers at Work: The Gender Division* (Garnaut and Lim-Applegate 1998; Garnaut *et al.* 1999). The second publication documented for the first time the average number of hours of work undertaken by men and women, both off-farm and on-farm. It also discussed the type of work carried out and the characteristics of Australian farm families by education, age and number of children. Garnaut *et al.* (1999) found that in 1994 women comprised 47% of the people in Australia's commercial farm businesses and 44% of Australia's livestock industries.

These reports documented that women also contribute to the industry as small farmers, new farmers, specialist operations, farm labourers, harvest operators, contractors, farm consultants, accountants, bankers, educators, trainers, researchers, media workers and government officers. Farmwomen were farm managers and administrators; they provided the farm 'help', as well as carrying out the more traditional roles of wife, mother, community worker and educator. Table 10.1 shows a clear gender divide in looking at the average hours per week spent in various roles by members of the family farm business. Predominantly males work in

Table 10.1. Average hours per week spent in various roles by members of the family farm business (broadacre and dairy industries, 1996–97).

	Females	Males
On-farm employment	19	51
Record keeping, etc.	5	3
Acquiring information and knowledge	1	3
Paddock and stock work and maintenance	13	45
Household work, childcare, community and voluntary work	40	7
Off-farm employment	7	5

(Source: Garnaut *et al.* 1999.)

the outside jobs of 'paddock and stock work and maintenance', and women in the inside domestic roles of child care and household work.

Research undertaken by Amabel Fulton and Cathy McGowan (2005) for Meat and Livestock Australia showed that any member of the farm family may act in one or more roles. These roles vary according to the family circumstances, the stage of farm succession, and the structure of the farm business. In some cases, women may be the major decision maker for particular aspects of the farm business. In other families, they may contribute to decisions. In some situations, they might make no overt contribution (Gasson and Errington 1993; Gooday 1995).

Barriers to participation

Mary Carroll, coordinator of the Irish women in agriculture program, summarised the barriers to women's participation as being the four Cs – culture, confidence, care and cash. Her observations ring true in the Australian environment. The Missed Opportunities report (Elix and Lambert 1998a, p. 92) identified the following barriers to women's participation:

- stereotyping/male attitudes
- own attitudes/self-confidence
- other commitments/interests/work/family
- lack of knowledge/skills/experience regarding the business
- lack of women in the industry/role models
- distance/isolation
- time
- access to/awareness of positions
- lack of money/business income
- age/physical ability
- staffing and supervision issues
- limited industry development.

As well as these barriers, there have been significant institutional barriers which have worked to exclude women and limit their participation in mainstream agricultural activities. Access to agricultural vocational education has been a key space where this exclusion was institutionalised. With no access to agricultural colleges, it was difficult for women to gain the formal education and hands-on knowledge and experience essential for farming. It was only relatively recently that the agricultural colleges, the main gateway into agricultural learning and jobs, allowed women to be students (Williams 1992).

While it was rare prior to the 1980s for women to undertake undergraduate and postgraduate courses in agricultural science, there have been exceptions. One notable woman scientist was Joan Tully who made an outstanding contribution to extension and research while working for the NSW Department of Agriculture, CSIRO and as a lecturer at the University of Queensland (Murray-Prior *et al.* 2006).

Until the mid-1970s, the majority of women with agricultural degrees focused their work in research and in academia. Few, to my knowledge, worked in mainstream agricultural extension and few regarded themselves as farmers. The acknowledged and accepted place for professional women working in the departments of agriculture was in the area of 'home economics'. Here, women with an interest in the home economy, in cooking, gardening and food, worked with farmers' wives.

The issue of access to education and training is not one confined to the last century. Fulton and McGowan's (2005) work with Meat and Livestock Australia and Australian Wool Innovation documented that women's participation in agriculture-specific learning programs in the meat and wool industries was still significantly below that of their male colleagues.

Another major institutional barrier to women's full participation in agricultural activities and decision making lies in the structure of farmer and agricultural organisations. A specific barrier is in the way voting rules are exercised. In most commodity organisations, voting rights are tied to levels of production and a certain percentage of production entitles the enterprise to a specific number of votes. In a farm family business it is rare for votes to be distributed to more than one person within the business. Usually it is the land owner, or main licence holder, i.e. the male farmer, whose name is registered and who gets to vote. As a consequence of these voting rules, in the past there were few women represented within farmer organisations.

Substantial advocacy work carried out by women's organisations in the 1990s brought about changes in many of the state-based farmer organisations. However, in the majority of the commodity groups, the rules around membership and voting continue to act as barrier to women's participation. Where organisations have changed their voting rules, significant increases in participation have occurred and gender representation is no longer an issue. In organisations such as the Queensland-based Agforce, the United Dairyfarmers of Victoria, and the Victorian Farmers Federation, there are now many examples of women taking senior leadership roles. With no vote, women's issues, and issues of young people are not seen as legitimate. This leads to a sense of being disenfranchised, and consequently a reluctance to participate in these organisations.

Clearly, women are not homogeneous. Women involved in family farm businesses reflect the diversity which exists within the general population. While at some level it is possible to form broad generalisations, particularly where there are children or older dependant family members to be cared for, there will be variations in age, in family lifecycles, in interest, and in levels of confidence. This diversity needs to be acknowledged and recognised with appropriate and well-researched strategies, some of which are discussed in the following section.

Strategies to increase involvement in extension activities

Unknowingly, the head of extension in making his comment '*We don't do gender, we do people*' highlighted an important issue for program designers. Perhaps his assumption was that by treating everyone the same, he would be able to avoid problems of discrimination. Alternatively, he may have thought that it would be too difficult, costly and time consuming to separate all the component target groups. Either way, his answer was problematic. The following section outlines some strategies that may help to increase the involvement of all farmers – i.e. all members of the family farm business – in extension activities.

(1) Understand the operating environment of the family farm business

In many instances, the farm is only one of a number of financial operations undertaken by the family. Income earned off-farm through investments or paid work all contribute to family income. To develop a clear understanding of the family as an economic unit, it is important to consider the financial operating environment of the farm in the context of all income-generating activities of the family, as well as understanding the goals, objectives, motivations and acknowledging the differing roles of the various family members in decision making about each economic activity and how they are interlinked. In many family farm businesses, income is brought into the family by the spouse working off-farm, for example as a teacher or nurse.

While they themselves may not be active in the farming business, it is often their income that plays an essential role in balancing the budget. Not only is this an issue of fairness and equality, spouses and members of the family farm business deserve the opportunity to be involved in and consulted about farm and business decisions because they have a valuable contribution to make.

In terms of participation in training programs and decision making, it is useful to include the appropriate and relevant people. However, it is not always easy to determine which members of the farm family team are responsible for which jobs. Sometimes farm families have never thought about this, or may feel inclined to hide the real work they do. Modesty and other forms of self-deprecation, as well as language issues and the way people respond in situations where powerful stereotypes operate, reduce the recognition given to the role of women on farms.

In her work investigating the gendered nature of work on Queensland sugar farms, Barbara Pini (2005) illustrated this lack of recognition. She nominated machines as being the main criterion for differentiating work that is designated male or female. Men's work is with machines. The cane-farming women who did engage in tractor work used a range of gender management strategies by which they sought to both undertake a masculine role and retain their femininity:

(1) They hid their on-farm contributions
(2) They emphasised the importance of their domestic and household roles
(3) They distanced themselves from men and by 'knowing and keeping their place and not trying to be one of the boys'
(4) They made sure they 'always acted like a lady'
(5) They referred to farming as a business and their role as being 'just another job'.

Women's contribution to the farm is partly determined by the stage of the family in the lifecycle. Using data from ABARE surveys of people in farming which examined the proportion of women in a range of farm roles based on their stage in the lifecycle, at least 72% of women are, at any one time, engaged in childcare of some form (Elix and Lambert 1998b).

This division of labour is given greater complexity depending on the stage of the family life cycle. Fulton and McGowan (2005) identified a typology of five indicative or archetypal caricatures (which were based on real case studies) which captured some of the diversity and mix of farm families in terms of life stage, children and farm business responsibilities.

(1) *The Assistant Farm Manager.* This type tends to be a 50 to 60 year old farming woman who has a grown family, some of whom may be part of the farm business. She works full-

time as a business/farm manager, occasional farm labourer and is generally responsible for farm administration and all family matters.

(2) *Off-farm income.* Typically this woman is a silent partner in the farm business. She is likely to be 40 to 50 years old. She may have no formal ownership of the farm or the business. Her major on-farm contribution is mothering and caring for older relatives and contributing to the family business income through off-farm work.

(3) *Business Manager.* The 40 plus year old woman is the business manager and farm partner, housewife and homemaker, typically with school-aged children. Her major contribution to the farm is her on-farm (but not farm-related) business and the income she makes which complements the farm income.

(4) *Mentor.* Woman aged 60 or more who work as a farm housewife and/or assistant farm manager. Her main role is supporting the extended farm families who work within the farm business. She is the grandmother, mentor to her daughters and sons-in-law and has occasional child care responsibilities.

(5) *Farm Labourer.* This woman is likely aged in her early 30s and is a mother and wife. She is responsible for the home and the children. Her major on-farm contribution is to 'help' on the farm with actual work and her off-farm income complements the family business income.

Each 'case study' offers different challenges and hurdles to be overcome and opportunities to be addressed when considering how best to involve women in extension programs. A number of consultants are now successfully working in this area, providing a specialist service to understand the situation of each family farm business.

(2) Consider time, place and venue

The location, timing and venue where activities are held have a surprising impact on the gendered participation of a gathering. To illustrate with an example from the 2003 drought and bushfire season, staff from the Department of Primary Industries ran a seminar on 'feeding stock in difficult times'. The venue was a local (male) farmer's shed which also enabled close proximity to the fodder and farm walk. The time of the session was from 7am to 11am, an early morning start including a BBQ breakfast. When the session was evaluated, the organisers commented on how disappointed they were with the small number of women who turned up.

Where were the women? Follow-up interviews identified the problem. The seminar was held at breakfast time, a peak period for many farm families. Mothers were doing family work. They were getting children ready for the day and ferrying them to the school bus. In making the family decision as to who should go to the seminar, the key information was that this workshop was about production. It was a busy time of day and only one member of the team could afford the time to attend. At that time of day, the least busy person was the male partner who also happened to have the management responsibility for production. In general, women's preference is for activities during school hours and for content to be aligned (targeted) to their particular farm roles.

(3) Horses for courses – gender-specific groups

Sometimes it makes sense to have a day or event targeted to a specific women's group. In the dairy, horticulture, sugar, cotton and grains industries, specific women's groups have been formed as a way of engaging with women in those industries. Initially funded by the respective Research & Development Corporations, the groups are now largely self-sustaining with active email discussion forums. The Wooragee Best Wool Best Lamb Group is one example. This learning group grew out of friendship links among local small-scale farmers in north-east Victoria. They meet monthly as part of a Meat and Livestock Australia and Australian Wool Innovation program. While the group is open to all, it specifically aims to encourage women to attend. When topics of interest to other members in the family team are up for discussion, they are also encouraged to attend. While targeted groups can be an effective extension technique, they work to best effect when they complement other extension activities and strategies. Women's groups work well for some women. Men's groups work well for some men. For others, mixed groups based on content related to specific skills are the preferred way of learning.

(4) Involve women in the planning – 'Nothing about us, without us'

A key strategy for effective research and extension is to ensure that specific processes are in place to involve, listen to, hear and implement the advice of women. This is particularly important in the design stages of an extension or change management program. Women have the knowledge about the operating environment and relevant issues. They have knowledge about available and appropriate venues, timing, speakers, and they understand managing transport and childcare issues. Through their local and regional networks, they also know how to attract other women.

For effective change and innovation, women and young people – in fact all members of the farm family team – need to be included at each level of decision making: in agricultural and rural policy making processes; in research and development programs; in change management processes; and in the evaluation and reporting mechanisms.

(5) Think about appropriate images and publicity

A picture speaks a thousand words. Traditionally, the public face of farming and agriculture has been the 'man on the land'. It is not unusual to pick up an agricultural publication, newspaper or annual report and find that all the images are of men, or where there are women, they are in 'helping' roles. A key strategy to increase involvement of more women and young people in programs is to have a diversity of images actively representing women, families, and young people in the marketing material, annual reports and rural and agricultural media. This can be particularly effective when local faces are used.

(6) Use appropriate language

Clarity of language is potentially one of the most useful strategies for working with women. When targeting women farmers, it makes sense to use the word 'women' in the title and in other places, for example: 'Women and men are encouraged to attend'. However, if the event is designed to cross a variety of 'jobs' and the 'right people' are being encouraged to attend, then clear language describing the roles and expected content needs to be used: e.g. 'This workshop will be particularly relevant to those in the business who are responsible for computer work.'

(7) Collect data by gender and age – disaggregate

When data are collected and analysed (disaggregated) by gender and age, they allow for decision making and targeting to be conducted more effectively. Alternatively, when data are grouped together and assigned to 'farmers', they can often be misleading. One example of this failure to disaggregate is the often quoted figure describing the education levels of 'Australian farmers'. When the phrase 'the average education of Australian farmers' is used, it describes the education level of 'the farmer' who filled in the census form. The statistics rarely reflect the education experience, qualifications and background of the farming women and children

Checklist for valuing women. (Source: Kerby et al. 1996, slightly modified.)

Changing the mind-set
Are you:
✓ Listening to the ideas, opinions and perceptions of both men & women?
✓ Recognising the varied skills, knowledge and experience of the clients?
✓ Acknowledging the different learning needs of men and women?
✓ Accommodating the learning needs of women and men?

Networking
Have you used networks by:
✓ Referring to your own list of women customers?
✓ Asking these women to invite other women?
✓ Sending the information to relevant organisations and groups?
✓ Accommodating the learning needs of women and men?

Inclusive language
Are you using language in which:
✓ People are treated equally?
✓ No irrelevance is introduced?
✓ No-one is excluded?
✓ The style is consistent?

Consultation and planning
In your planning:
✓ Are women involved in the process?
✓ Do women have some ownership of the activity?
✓ Does the activity address the priorities of all clients including women?

In designing the program:
✓ Have you discussed with women an appropriate format, venue and content for them?
✓ Do they prefer separate activities (to men) or combined?
✓ Do the speakers/contributors you have engaged reflect the input of the planning group?

✓ Is it possible to include women as speakers or contributors?
✓ Have you briefed your speakers about inviting participation from the women present?

Planning an activity
Is the venue:
✓ Appropriate for women?
✓ Known to women?
✓ Comfortable and inviting?
✓ Accessible?
✓ Used by them for other activities?
✓ Offering appropriate facilities for women and men?
✓ Are the timing, costs and childcare appropriate for women?

Childcare arrangements
In organising childcare:
✓ Have you included the costs in your administration costs?
✓ Does the venue have childcare facilities?
✓ Have you encouraged the attendance of children if childcare is not available and provided activities for them in the program?
✓ Have you offered to reimburse the parents for the cost of childcare?

Promotion
In your promotion, have you:
✓ Targeted women?
✓ Direct mailed both partners?
✓ Personally invited women by phone, word of mouth or letter?
✓ Informed community and school newsletters?
✓ Used daytime television and radio?

Evaluation
In evaluating the activity:
✓ How will you measure the outcomes of the activity against the objectives?
✓ Have you allowed for feedback from participants?

who also contribute to the family farm business and frequently have higher qualifications than 'the farmer'.

(8) Use Kerby's checklist

Kerby *et al.* (1996) translated the barriers to women's participation in agricultural industry programs into a checklist for designing, implementing and evaluating programs. Where this checklist has been used (for example in programs in the dairy, horticulture and sugar industries), participation of women has increased.

Epilogue

Improving irrigation efficiency was the topic for a farm extension day. The one-day seminar was designed to bring innovations in farming practice to local farmers. The topic was important especially to Jenny and John Rice, who worked in a farming partnership. They worked well together and enjoyed farming and their family lifestyle. Jenny earned occasional off-farm income and was responsible for all the administration, computer work and financial records of their business. Their eldest son, Jack, was spending his gap year at home working on the farm.

John Rice was keen to attend the seminar and there wasn't much discussion in the family about who was going to attend. At dinner after the field day, John was excited by what he had learnt. It was *'a very good day, lots of farmers'*, and he shared with Jenny and Jack what he had heard about the new mechanisms for testing soil moisture and for managing plant growth.

The next day Jenny met a woman farming friend in the supermarket. *'Hi Jenny, you missed a great session on irrigation yesterday.'* *'Yes'* said Jenny, *'John said it was good. He told us all about it at dinner last night.'* *'That's great. Then you will be getting the new computer program that links the watering systems to the accounting package'* said Jane. *'Oh, he didn't mention computers'*, said Jenny, seething. She hated it when John paid no attention to the computers and finances, leaving it totally to her. Here, she felt, was just another example where he failed to listen.

Jack was a keen football player. At practice on Wednesday night, his mates were also discussing the field day. *'It's great how they can send all the information by SMS,'* they enthused. *'What's that?'* Jack asked. *'Well, they can use the mobile to text prices for water, as well as times for releasing your allocation.'* Jack gave a deep sigh, another example of his parents being totally out of date. Why hadn't his Dad told him about this? It would make a big difference to the amount of time he spent organising water. But he knew why. His Dad didn't know the first thing about SMS, and with those thick fingers of his, he wasn't going to start learning now.

At dinner that evening, there was an interesting family discussion. Where did the problem lie? Was it with John for not picking up the information that he knew would have been of interest and relevance to the jobs of his family? Was it with Jenny and Jack for failing to appreciate that the day was about all the skills needed on the farm and that they too should have gone to the seminar? What about the people who designed the seminar publicity? If they had given more thought to the intended audience and included details about the content, expectations and outcomes of the day, would it have helped the family decide who should attend?

Jenny and Jack agreed that if they had known what the day was to have been about, they all would have made the time to attend and they would have enjoyed the quality family time in the car travelling to and from the seminar.

Acknowledgements

Many of the ideas in this chapter have been developed in partnership with colleagues working in the field. In particular, I would like to acknowledge the contribution of Dr Amabel Fulton in sharing her analysis of women's participation in the Australian meat, sheep and horticultural industries.

References

Dempsey K (1992) *A Man's Town: Inequality between Women and Men in Rural Australia.* Oxford University Press: Melbourne.

Elix J and Lambert J (1998a) *Missed Opportunities: Harnessing the Potential of Women in Australian Agriculture, Volume 1: Social Survey and Analysis.* Rural Industries Research and Development Corporation and Department of Primary Industries and Energy: Canberra.

Elix J and Lambert J (1998b) *Missed Opportunities: Harnessing the Potential of Women in Australian Agriculture, Volume 2. Economic Module: Estimating the Current and Potential Contribution of Women in Agriculture.* Rural Industries Research and Development Corporation and Department of Primary Industries and Energy: Canberra.

Fulton A and McGowan C (2005) *Fostering Women's Participation in On-Farm Programs.* Meat and Livestock Australia Limited: Sydney.

Garnaut J and Lim-Applegate H (1998) 'People in farming'. ABARE Research Report 98.6. ABARE: Canberra.

Garnaut J, Rasheed C and Rodriguez G (1999) 'Farmers at work: the gender division'. ABARE Research Report 99.1. ABARE: Canberra.

Gasson R and Errington A (1993) *The Farm Family Business.* CAB International: Wallingford.

Gooday J (1995) 'Women on farms: a survey of women on Australian broadacre and dairy farms, 1993–1994'. ABARE Research Report 95.10. ABARE: Canberra.

Kerby J, McClure L, Westley A and Young M (1996) 'Valuing women as customers'. Technical Report No. 245. Primary Industries South Australia: Adelaide.

Kilpatrick S and Bell R (2000) Sharing the driving seat: involving everyone in a family business. *Rural Society* **10**(1), 5–14.

Moi T (2002) *Sexual/Textual Politics: Feminist Literary Theory.* 2nd edn. Routledge: New York.

Murray-Prior R, Jennings J and King C (2006) Our Australian idol – Joan Tully. *APEN Newsletter* **14**(2), 1–2.

Pini B (2005) Driving tractors and negotiating gender. *International Journal of Sociology of Agriculture and Food* **13**(1), 1–18.

Silvasti T (2003a) Bending borders of gendered labour division on farms: the case of Finland. *Sociologia Ruralis* **43**(2), 154–166.

Silvasti T (2003b) The cultural model of 'the good farmer' and the environmental question in Finland. *Agriculture and Human Values* **20**(2), 143–150.

Vanclay F (2003) The impacts of deregulation and agricultural restructuring for rural Australia. *Australian Journal of Social Issues* **38**(1), 81–94.

Vanclay F, Silvasti T and Howden P (2007) Styles, parables and scripts: diversity and conformity in Australian and Finnish agriculture. *Rural Society* **17**(1), 3–18.

Waring M (1988) *Counting for Nothing: What Men Value and What Women Are Worth.* Allen & Unwin: Sydney.

Williams J (1992) 'The invisible farmer: a summary report on Australian farm women'. Commonwealth Department of Primary Industries and Energy: Canberra.

11

Bridging the gap between policy and management of natural resources

Allan Curtis and Emily Mendham

Summary

This chapter provides insights for regional natural resource management (NRM) practitioners seeking to influence property management by rural landholders. We aim to bridge the gap between policy and management by providing improved understanding of changes in the social structure of rural areas and the impact of these changes and other influences on landholder decisions. We draw upon a number of theoretical frameworks, including the adoption literature. We also draw upon a substantial body of empirical research exploring the social structure of rural areas and landholder implementation of sustainable farming and biodiversity practices, including recent studies undertaken in the Corangamite and Wimmera regions of Victoria. Our research was intended to help regional NRM practitioners engage rural landholders, develop an effective mix of policy approaches, and evaluate the accomplishment of intermediate NRM program objectives. Key findings include: (1) only part of the implementation of conservation activities by landholders can be directly attributed to investment by NRM programs; (2) investing in NRM programs that engage human and social capital is an effective way of influencing the property management of landholders; and (3) NRM practitioners need to be aware that they will be engaging a very different cohort of rural landholders than in the past, with significantly increased proportions of non-farmers and absentee property owners.

Introduction

NRM in Australia is increasingly structured around a regional delivery model where planning and implementation is guided by national and state priorities. In such a geographically large country, there are compelling reasons for managing at a bio-regional level. The regional scale is considered the most appropriate to support NRM that is holistic (covers all landscape elements); systematic (considers the interactions between elements); and comprehensive (embraces the range of values attached to landscapes) (Bammer *et al.* 2005). The regional scale also has attractions for policy and program managers concerned about the efficient and effective management of public funds (Curtis and Lockwood 2000).

Given the scale of land and water degradation and the relatively small tax base from which to fund remediation, Australians have relied heavily on the voluntary work of around 120 000 farming families who manage most of the continent. In some instances, regional NRM practitioners deliberately seek to engage individual landholders in practice change. One example would be where a landholder manages all or part of a critical environmental asset. As Pannell *et al.* (2006) and the authors of other chapters explain, engaging landholders in practice change is complex and difficult, not least because there is a potentially large set of factors influencing decisions and these vary according to each technology and landholder context. Even the concept of implementation is problematic. For example, when does a trial change in practice represent implementation? In many instances, personal engagement with individuals by regional NRM practitioners will not be feasible or even necessary. However, it should be possible for regional NRM practitioners to develop a suite of policy instruments that meet the diverse needs of landholders and are adapted over time. An effective suite of policy instruments might include one-to-one or group extension, offering to share the cost of work on private land; and regulation that ensures landholders meet a duty of care for the environment (Dovers and Mobbs 1997).

Despite the inherent appeal of regionalism, the 56 regional NRM bodies established in the Australian states since the early 1990s have faced a myriad of challenges (Paton *et al.* 2004). As the experience with watershed organisations in the United States confirms, regional NRM practitioners will find it difficult to engage landholders who typically identify and operate at the local scale (Curtis *et al.* 2005). Regional NRM practitioners have also struggled with the task of evaluating the outcomes of investments made through regional plans (Paton *et al.* 2004). Insufficient attention by federal and state agencies to the monitoring of trends in resource condition is part of the explanation of the strong critique delivered by successive Australian National Audit Office reviews of major NRM programs (ANAO 1997, 1998, 2001, 2008). The challenges of landholder engagement, implementation and evaluation are related. One way to measure program effectiveness is by using intermediate measures of program objectives, including the implementation of practices expected to lead to improved resource condition outcomes (Curtis *et al.* 1998).

The research findings presented in this chapter should provide the knowledge base to underpin improved performance by regional NRM practitioners as they engage rural landholders, develop a suite of policy instruments and evaluate NRM program outcomes. We draw upon a substantial body of empirical research where the lead author has explored the social structure of rural areas and landholder implementation of sustainable farming and biodiversity conservation practices. Instead of relying on the analysis of secondary data, this research has collected and analysed large samples of spatially referenced survey data provided directly by rural landholders. The most recent studies were undertaken in the Corangamite and Wimmera regions of Victoria (Curtis *et al.* 2006, 2008b). With surveys in 2002 (Curtis and Byron 2002) and 2007 (Curtis *et al.* 2008b) in the Wimmera region, we are able to draw upon the only longitudinal study of this type in an Australian region. Our view is that the two regions we studied provide a robust snapshot of the majority of the Victorian Catchment Management Authority (CMA) regions, with the possible exception of the irrigated areas along the Murray.

In the next section, we explain our survey approach, including our reasons for collecting spatially referenced data. We present our key findings including the importance of voluntary work by landholders, the efficacy of programs that engage and build human and social capital, and the challenge being faced by many NRM practitioners who are beginning to work with a significantly different cohort of rural landholders than previously.

Background

Our experience working with regional organisations suggests that social researchers can make important contributions to the knowledge base that underpins regional NRM. The groups we have worked with said that their first priority was for social research that would assist them to:

(1)   identify and refine investment priorities
(2)   develop and improve engagement with private landholders
(3)   choose from amongst the mix of policy instruments available to accomplish resource condition targets
(4)   evaluate the achievement of intermediate NRM objectives over time.

Of course, there are many ways to accomplish these tasks (Cavaye 2003; Curtis *et al.* 2005). The analysis of data collected through farm and household censuses can provide useful information, particularly about trends in the social structure of regions (Barr *et al.* 2000). However, as Shultz *et al.* (1998) and Curtis *et al.* (2001) have demonstrated, these data are unlikely to satisfy regional NRM practitioners who need to understand the factors influencing property management by private landholders. In the first instance, these national data collection processes are unlikely to address most of the topics for which data are needed. The second major limitation is that data are typically only available at aggregates of 200 households or so for each local government area. It is impossible to explore the factors affecting individual landholder implementation with these aggregated datasets.

With limited resources to fund social research, our regional partners chose to focus on gathering spatially referenced socio-economic data using a mail survey to rural landholders that would enable them to address each of the four topics identified above. Drawing on our experience working with regional groups in Victoria, Queensland and New South Wales from 1998 to 2002 (e.g. Byron and Curtis 2002a, 2002b; Curtis *et al.* 2000); the extensive literature on adoption studies in Australia (e.g. Barr and Cary 1992; Cary *et al.* 2002; Vanclay 1992, 2004); our studies of adoption of sustainable farming practices (Curtis and De Lacy 1996), native grasses (Millar and Curtis 1997), farm forestry (Race and Curtis 1998), riparian areas (Curtis and Robertson 2003a) and native vegetation (Mendham *et al.* 2007); and our understanding of program evaluation theory (Cook and Shadish 1986; Patton 1990) and experience with NRM program evaluations (Curtis *et al.* 1998); we identified and refined a set of topics that would provide information to underpin key elements of the work of regional practitioners. This set of topics included each landholder's:

- level of awareness and concern about the social, economic and environmental issues affecting their property and region
- values which they attach to their property
- knowledge and understanding of processes leading to land and water degradation and of how to implement practices expected to lead to improved resource condition outcomes
- level of confidence in recommended practices
- property size
- on- and off-property income and debt levels
- involvement in short courses
- stage of life and involvement in succession planning
- long-term plans for their property
- information sources
- current and future land use
- implementation of practices expected to lead to improved resource condition outcomes
- access to government financial support for implementation
- responses to policy instruments expected to change landholder behaviour and achieve regional targets.

The Corangamite NRM region (13 340 km²) is located to the west of the city of Melbourne in the state of Victoria, Australia (Figure 11.1). The region has a diverse economy with

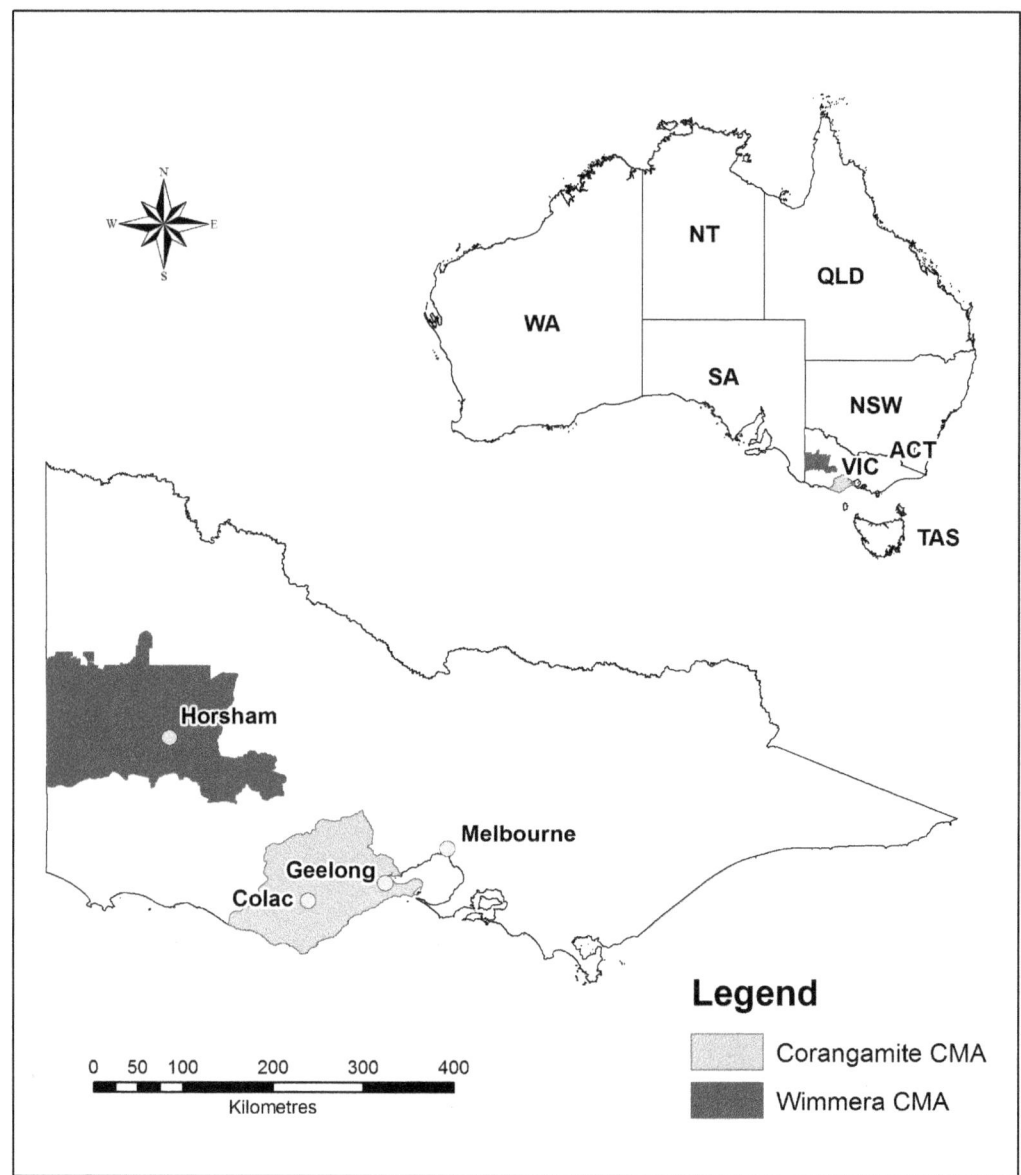

Figure 11.1. Location of Corangamite and Wimmera regions in south-eastern Australia.

employment dominated by the manufacturing and services sectors. Land is principally used for agriculture and includes livestock grazing (beef and dairy) and cropping, although agriculture only employs 5% of the population (Corangamite Catchment Management Authority 2003). Access to and from the region to metropolitan Melbourne and the adjoining city of Geelong is enhanced by a fast train route and a high quality road network. Many parts of the region are within a one-hour commute of Melbourne.

The Wimmera catchment (23 500 km²) is located in western Victoria, north-west of Corangamite (Figure 11.1). The region has a diverse environment, including open forests, agricultural landscapes, the Wimmera River, wetlands, lakes and intermittent streams. The main land use of the region is broadacre agriculture, with approximately 85% of the region cleared of native vegetation. The population of about 50 000 lives mainly in the regional centres of Horsham and Stawell, as well as smaller townships.

Corangamite and Wimmera were selected from the five Victorian regions where our research has been completed because they are the most recent surveys and together provide a robust representation or 'slice' of Victoria (save for Murray irrigation areas). These regions are representative of most of regional Victoria in that they include major cities (Geelong, Colac and Horsham); inland and coastal areas; extensive dryland and irrigation agriculture; areas of high amenity value (the Otways and Grampians National Parks) and areas largely focused on production. Both regions have a mix of landholders with a full-time focus on production and an increasing influx of 'hobby farmers'. Selecting two regions allows us to compare trends occurring across Victoria without attempting to present more data than space permits or is needed to explain our key findings.

Local governments in the two regions provided access to ratepayer lists and these were used to compile a list of all rural properties greater than 10 hectares (used to separate rural and urban land use). These lists included a property identification field that supported spatial referencing of the survey data. The survey design and mail-out employed a modified Dillman (1978) *Total Design Method* process that has been refined through the experience of successive catchment surveys. Curtis *et al.* (2005) provides a detailed explanation of the collaborative research process undertaken. In summary, a 12-page survey booklet was developed in collaboration with regional partners, extensively pre-tested through workshops with landholders, and mailed to selected respondents. In the Wimmera 2007 survey, a list of 1200 landholders was randomly selected from ratepayer lists provided by local governments. After removing multiple listings of properties and known deceased estates, the final mailing list contained 1000 landholders. The final response rate was 56% from this survey (useable n=503). In 2002, surveys were sent to 959 landholders in the Wimmera region from a random sample of 1000 selected from local government lists. The final response rate was 73% (useable n=619). In the Corangamite 2006 survey, the final mailing list contained 972 landholders. The final response rate was 57% (useable n=482).

The survey topics in the Wimmera and Corangamite projects were consistent with those described earlier, but were modified to reflect the context in each region. The practices expected to lead to improved resource condition included in the surveys were slightly different in the two regions (Table 11.1). Survey respondents were asked about their implementation of Current Recommended Practices (CRP) for both sustainable agriculture and biodiversity conservation. CRP included in the surveys were identified by Catchment Management Authority (CMA) staff or by participants in the survey pre-testing workshops as those practices expected to lead to improvements in catchment condition. CRP included in the survey can be classified into two groups: those principally related to biodiversity conservation (such as area of trees/shrubs planted); and those related to sustainable agriculture (such as soil tests).

We were conscious that some CRP are relevant to most landholders (i.e. non-specific, such as tree planting) while others are more relevant to particular landholders (i.e. specific, such as implementing minimum tillage in cropping systems). For this research, we identified CRP specific to either cropping (such as direct seeding) or livestock grazing (fencing bush to limit stock access). All respondents were included in calculations of the percentage of respondents implementing non-specific CRP, but only those involved in cropping or livestock enterprises were included in calculations for the proportion of respondents implementing CRP specific to each type of enterprise, or in analyses exploring the factors linked to implementation of CRP. As shown in Table 11.2, specific survey items were included to explore implementation over the

Table 11.1. Current Recommended Practices (CRP) included in the Wimmera 2007 and Corangamite 2006 surveys.

Current Recommended Practice (CRP) survey questions	Wimmera	Corangamite
Area of trees and shrubs planted/direct seeded	✓	✓
Area of farm forestry established	✓	✓
Length of fencing erected to manage stock access to rivers/streams/wetlands	✓	✓
Area of native bush/grasslands fenced to manage stock access	✓	✓
Area sown to perennial pasture and lucerne	✓	✓
Maximum area of crop (of any type) sown in any year using minimum tillage techniques	✓	✓
Have you tested the water quality of the main water source for stock or irrigation purposes	✓	✓
Number of off-stream watering points established	✓	✗
Area of gully erosion addressed	✓	✗
Maximum area of crop sown in any year using no-till techniques	✓	✗
Number of paddocks for which you have a record of soil test results	✗	✓
Area with at least one lime application	✗	✓
Area cropped using a rotation with pasture (e.g. lucerne)	✗	✓
Area where you used time controlled or rotational grazing	✗	✓
Estimated time spent by you or others to control pest animals and non-crop weeds	✗	✓

full period of a respondent's property management and others explored implementation over much shorter timeframes, mostly the last five years. Seven CRP were common to both surveys.

Findings

Nurture the voluntary work of landholders

Challenges for NRM practitioners

The Corangamite survey explored the implementation of 12 CRP through 12 items (Table 11.2) and the Wimmera survey 10 CRP through 15 items (Table 11.3). Seven items were common to both surveys. Overall, there were similar levels of implementation in both regions in that more than 50% of respondents said they had implemented about half of the CRP in each region (six of 12 CRP in Corangamite and four of 10 CRP in Wimmera). However, there were only two items, both for Corangamite, where more than 60% of respondents had implemented the specific CRP. Thus, in both regions the level of adoption is lower than might be seen as desirable. It seems there are considerable challenges for NRM practitioners seeking to engage landholders in these practices.

Importance of government funding

Just over half of the Wimmera respondents (56%) said that work undertaken to implement at least one of the CRP had been supported by the financial or technical resources provided by

Table 11.2.   Implementation of CRP, Corangamite 2006 (n=482).

Corangamite 2006: Practices undertaken during your management	n	% adopt	median
Area of trees and shrubs planted (incl. direct seeding) (ha)	445	75	4 ha
Area of farm forestry established (ha)	449	17	5 ha
Area of native bush/grasslands fenced to manage stock access (ha)	436	31	5.5 ha
Length of fencing erected to manage stock access to rivers/streams/wetlands (km)	444	49	3 km
Have you tested the water quality of the main water source for stock or irrigation purposes on your property in the last five years?	461	27	N/A
Number of paddocks where have a record of soil test results for past five years	435	55	6
Area sown to perennial pasture and lucerne (ha) past five years	451	52	40 ha
Area cropped in past five years using a rotation with pasture e.g. lucerne	446	49	50 ha
Area where used time controlled rotational grazing in past 12 months	448	52	97.5 ha
Area with at least one lime application past five years	449	54%	50 ha
Max area of crop sown in any year using min-tillage techniques (ha) past 12 months	450	42	38.5 ha
Time spent by you/others to control pest animals and weeds in past 12 months	449	88	10 days

government, including work that was then implemented by the Wimmera CMA, local landcare group or networks, state government agencies or non-government organisations. Thirty-six per cent of Corangamite respondents said government funding had supported work on their property in the past five years. Our analyses established the existence of significant positive relationships between government funding and the implementation of seven out of the 10 CRP in the Wimmera study (Curtis *et al.* 2008b) and all 12 CRP in the Corangamite study (Curtis *et al.* 2006), providing a strong evidence base to suggest that government support makes a difference to implementation.

More to the story: the importance of voluntary contributions

The remainder of this section focuses on the Wimmera region where we have longitudinal data and information about the support from government for implementation of specific CRP. A key finding is that there was no item where more than 50% of those implementing a practice said they had been supported by government (Table 11.4). In other words, most of those implementing work had done so without direct government support.

Evaluating intermediate NRM outcomes

An important aim of the survey was to assist regional practitioners to evaluate the impact of their NRM programs. As social researchers, we worked with our regional partners to identify intermediate program objectives that were supported by theory and empirical evidence. Typically, these intermediate objectives included landholder awareness of issues, knowledge of degradation processes and practices expected to address these threats, confidence in CRP and

Table 11.3.   Implementation of CRP, Wimmera 2007 (n=503).

Wimmera 2007: Practices undertaken during your management	n	% adopt	median
Area of trees and shrubs planted (incl. direct seeding) (ha)	487	54	5 ha
Area sown to perennial pasture and lucerne (ha)	490	36	75 ha
Area of gully erosion addressed (ha)	487	11	5 ha
Area of farm forestry established (ha)	489	10	5 ha
Area of native bush/grasslands fenced to manage stock access (ha)	393	37	10 ha
Length of fencing erected to manage stock access to rivers/streams/wetlands (km)	392	27	4 km
Number of off-stream watering points established (number)	393	23	5.5
Have you tested the water quality of the main water source for stock or irrigation purposes on your property in the last 5 years?	428	58	N/A
Area of trees and shrubs planted (incl. direct seeding) (planted annually) past 5 years	491	37	200 *
Area sown to perennial pasture and lucerne (ha) past five years	489	30	60 ha
Area of farm forestry established (ha) past five years	491	4	5 ha
Length of fencing erected to manage stock access to rivers/streams/wetlands (km) past five years	395	21	3 km
Area of native bush/grasslands fenced to manage stock access (ha enclosed annually) past five years	490	20	2 ha
Maximum area of crop sown in any year using no-till techniques (ha) past five years	32	56	200 ha
Max area of crop sown in any year using min-tillage techniques (ha) past five years (during past 12 months Corangamite)	33	52	200 ha

* Data provided in survey recalculated as number of trees per year

implementation of CRP. In each of our studies, there has been a close alignment between the targets identified in regional catchment strategies and the intermediate objectives included in our survey. The items included in both the 2002 and 2007 Wimmera surveys enabled a comparison of trends over time. NRM investment is increasingly targeted to specific asset classes, such as a vegetation type or a specific wetland. With spatially-referenced survey data we were able to test for changes over time in intermediate objectives for all respondents and for those in locations with specific assets. We discuss findings from the regional-scale analysis first. In presenting key findings, we focus on the implementation of on-ground work.

Comparison of the 2002 and 2007 survey data suggests that at the regional scale there has been a general increase in awareness of river health, water quality, dryland salinity and soil erosion issues, as well as an increased preparedness of landholders to acknowledge the impact of their land use on soils (Curtis *et al.* 2008b). Two topics exploring landholder confidence in CRP were included in both surveys and there was evidence of significantly increased levels of confidence in these items: fencing to control stock access as an essential part of work to revegetate waterways; and watering stock off-stream to improve bank stability, water quality and

Table 11.4.   Implementation of CRP and government support, Wimmera 2007 (n=503).

	n	% adopt	work median	% receiving govt support
Practices undertaken during your management				
Area of trees and shrubs planted (including direct seeding) (ha)	487	54	5 ha	44
Area sown to perennial pasture and lucerne (ha)	490	36	75 ha	10
Area of gully erosion addressed (ha)	487	11	5 ha	16
Area of farm forestry established (ha)	489	10	5 ha	10
Area of native bush/grasslands fenced to manage stock access (ha)	393	37	10 ha	32
Length of fencing erected to manage stock access to rivers/streams/wetlands (km)	392	27	4 km	33
Number of off-stream watering points established (number)	393	23	5.5	6
Practices undertaken in last five years				
Area of trees and shrubs planted (including direct seeding) (ha)	491	37	4 ha	32
Area sown to perennial pasture and lucerne (ha)	489	30	60 ha	12
Length of fencing erected to manage stock access to rivers/streams/wetlands (km)	395	21	3 km	23
Area of native bush/grasslands fenced to manage stock access (ha)	490	20	10 ha	20

stock condition (Curtis *et al.* 2008b). Conversely, there were significantly lower self-reported levels of knowledge for nine of the 12 topics included in both surveys. This trend remained for analyses focused on the eight knowledge topics where the Wimmera CMA has targeted investments on specific assets. Indeed, there had been a significant decline for six of the eight topics (Curtis *et al.* 2008b). In summary, although landholder awareness and confidence in CRP appeared to be increasing, landholder knowledge on topics where the CMA had targeted investment was declining.

Assessing implementation of CRP

It was possible to compare 2002 and 2007 data for six items relating to five of the 10 CRP (Table 11.5). Significantly lower proportions of landholders in 2007 were involved in three CRP (trees and shrubs planted past five years; perennial pasture established during management period; cropping using minimum tillage past five years), significantly increased proportions involved in one CRP (farm forestry), and there was no clear trend for the remaining CRP (fencing to manage stock access to native bush or grasslands).

Calculations of the average (median) amount of work completed by respondents for the five CRP where comparisons could be made provided additional information for assessing the achievement of on-ground objectives. Although fewer respondents had established trees and shrubs, there was a significant increase in the median number of trees and shrubs established. For all other CRP, the median amount of work declined over time (significantly for perennial

Table 11.5.   Implementation of CRP, Wimmera 2007 (n=503) and 2002 (n=619).

Practices undertaken	Survey year	% adopt	p-value	Work done median	p-value
Area sown to perennial pasture and lucerne during period of management (ha)	2007 (n=490)	36% (n=174)	0.0001	75 ha	0.0011
	2002 (n=590)	47% (n=279)		120 ha	
Area of native bush/ grasslands fenced to manage stock access during period of management (ha)	2007 (n=393)	37% (n=145)	0.0776	10 ha	0.022
	2002 (n=292)	30% (n=88)		20 ha	
Area of farm forestry established during period of management (ha)	2007 (n=489)	10% (n=36)	0.0439	5.5 ha	0.411
	2002 (n=587)	6% (n=28)		10 ha	
Area of native bush/ grasslands fenced to manage stock access during the past five years (2007) or past three years (2002) (ha enclosed annually)	2007 (n=195)	32% (n=54)	0.1803	0.75 ha	0
	2002 (n=416)	27% (n=139)		5 ha	
Area of trees and shrubs planted (incl. direct seeding) during the past five years (2007) or past three years (2002) (planted annually)	2007 (n=491)	37% (n=183)	0	200 *	0
	2002 (n=580)	60% (n=346)		83.3	
Max area of crop sown in any year using minimum tillage techniques during the past five years (2007) or past three years (2002) (ha)	2007 (n=351)	66% (n=224)	0.0002	300 ha	0.1959
	2002 (n=471)	77% (n=271)		360 ha	

Note: *Area recalculated as number of trees per year

pasture established; fencing to manage stock access to native bush/grasslands) (Table 11.5). The trend to lower levels of implementation remained for analyses focused on specific assets (Table 11.6) in that there was a trend to lower levels of landholder involvement in four of the five CRP (with the exception of fencing to manage stock access to native bush/grasslands), with a statistically significant decline for two CRP (trees and shrubs planted; cropping using minimum tillage).

In our discussion of preliminary research findings with Wimmera CMA staff, it became clear that the CMA was accomplishing its targets for work that was directly funded and managed by CMA staff or contractors. However, our findings indicated that it was possible to achieve the targets for direct CMA investment in on-ground work, but go backwards in terms of the proportion of landholders implementing CRP and the median amount of work implemented by landholders across the region.

Table 11.6.   Implementation of CRP by strategic investment areas, Wimmera 2002 (n=619) and 2007 (n=503).

CRP	Investment asset/area	% involved		p-value
		2002	2007	
Area of trees and shrubs planted (including direct seeding) last five years	Three priority ground flow systems	48% n=33	43% n=51	0.7976
	Yarriambiack and Hindmarsh	56% n=282	32% n=220	0.000
Area sown to perennial pasture and lucerne	Three priority ground flow systems	64% n=33	45% n=51	0.1405
Max area of crop sown in any year using minimum tillage techniques	Wimmera cropping areas	77% n=471	66% n=351	0.002
Area of farm forestry established	Three priority ground flow systems	15% n=33	4% n=50	0.1659
Area of native bush/grasslands fenced to manage stock access (ha)	Wimmera region	30% n=88	37% n=145	0.0776

Other survey findings provide at least part of the explanation for the trend to reduced involvement of landholders in implementation of CRP. There has been a severe drought in the Wimmera for most of the period since 2002. Survey data shows that median on-property income dropped from $45 000 to $15 000 per annum, and that only 35% of respondents made a profit in 2006/07 compared to 86% in 2002. Drought conditions might also have made it impractical to implement some practices and resulted in farmers or their partners moving off-property for work. Indeed, just over three-quarters of the survey respondents said they or their partner had received a net off-property income in 2006/07, up from 66% in 2002. Again, there is a potential for this trend to reduce the capacity of landholders to implement on-ground work. This hypothesis appeared to be supported by findings that on-property work was positively linked to implementation of CRP while off-property income was negatively linked to implementation. Given that most of the work implemented by Wimmera survey respondents was self-funded and that in 2007 on-property profitability was linked to significantly higher implementation of six CRP, it seems likely that the impact of drought is at least part of the explanation for lower levels of implementation across the Wimmera between 2002 and 2007 (Curtis and Byron 2002; Curtis *et al.* 2008b).

Build and engage human and social capital

Introduction
The 'capitals' concept is a widely accepted and useful framework (Australian Bureau of Statistics 2002; Black and Hughes 2001) and can be applied to the task of understanding the complex web of factors that affect landholder capacity (i.e. ability) to implement recommended practices (Webb and Curtis 2002; Pannell *et al.* 2006). The concept of capital recognises that there is a stock of resources that can be used to achieve some desired endpoint such as improved NRM. Economic resources and physical infrastructure are types of human-created capital.

The skills, abilities and wellbeing of the population form our human capital, while social capital refers to the social relations, networks, trust and norms that arise between people when they interact and which can then lead to further benefits. Natural capital refers to the non-human parts of the world upon which human existence is premised (Sobels *et al.* 2001; Castle 2002; Webb and Curtis 2002).

As social researchers, our focus has been on exploring the assumed relationships between investments that engage or build human and social capital and landholder practice change. We use the terms 'engage' and 'build' advisedly. It is our view that researchers, and to a lesser extent NRM practitioners, often underestimate the existing capacity of landholders.

Through our research, including the Corangamite and Wimmera regional landholder surveys, we have explored relationships between landholder capacity, the policy instruments available to regional NRM practitioners, and implementation of CRP by landholders. Our work builds on an extensive body of Australian research examining links between human capital and landholder practice change (Barr and Cary 2000; Cary *et al.* 2002; Guerin and Guerin 1994). The topics included in our surveys have therefore been shaped by the findings of past research, including our own (e.g. Curtis and De Lacy 1996; Curtis and Robertson 2003b) and the imperative of focusing on variables that can be influenced by regional NRM practitioners, or at least will provide useful information to inform their engagement with landholders.

Before moving to a discussion of key findings from the Corangamite and Wimmera surveys, it is important to address concerns that have been raised about the value of capacity building investments. In a recent review for the Department of Sustainability and Environment, Curtis *et al.* (2008a) drew on the Victorian experience with landcare to examine the 'value proposition' for government investment in approaches that engage and build human and social capital to achieve NRM outcomes. The authors drew on rural development and extension theory and empirical research evaluating landcare (see Curtis 2007) to articulate an evaluation framework and marshall the available evidence, mostly from Victoria. The authors concluded that there was substantial evidence suggesting that landcare has had a significant positive impact on human and social capital and that these changes had contributed significantly to practice change. The authors also acknowledged the limitations of capacity building programs and the importance of a mix of policy approaches to NRM.

Capacity building enhances NRM outcomes

Findings from the recent Corangamite and Wimmera studies provide considerable evidence of the positive impact of investments in capacity building – including through short courses such as those related to property management, landcare and commodity groups – that focuses on raising awareness, improving knowledge and building confidence in CRP (Table 11.6). These capacity building platforms, and others, are typically supported by direct government investment. It is also important to acknowledge that in most instances these platforms would not exist or succeed without substantial input from landholders and other volunteers. Because direct government investment can occur outside these platforms, we have included government funding received as a variable in the list of other factors that practitioners should consider when seeking to engage landholders in practice change. As explained, the Corangamite survey explored the implementation of 12 CRP and the Wimmera survey 10 CRP (Table 11.2). All relationships reported in Tables 11.7, 11.8, 11.9 and 11.10 were statistically significant at the 95% confidence level using a variety of pair-wise tests, including the Kruskal-Wallis Rank Sum Test, Spearman rank order correlations and the proportions test. Given the weight of evidence provided by the large number of statistically significant relationships summarised in Tables 11.7 and 11.8, we have not included findings from multivariate analyses.

Table 11.7.   Significant positive relationships between capacity building approaches and implementation of CRP, Corangamite 2006 (n=482) and Wimmera 2007 (n=503).

Capacity building topics	Corangamite n=482 (12 CRP)	Wimmera n=503 (10 CRP)
Awareness of salinity on property	10 practices (all except trees and shrubs planted, used time controlled/rotational grazing) and said areas property affected by dryland salinity	Three practices (trees and shrubs planted, gully erosion control, sowing perennial pasture) and said areas property affected by dryland salinity
Self-assessed knowledge of NRM	12 practices and 11 of 17 landholder knowledge items	Seven practices (all except testing water quality, establishing off-stream watering, gully erosion control) and all 17 landholder knowledge items
Confidence in CRP (Survey items focused on fencing to manage stock and tillage practices)	Two practices (fencing waterways, fencing native vegetation) and confidence in benefits of managing stock access to waterways and native vegetation	One practice (min-tillage) and confidence in benefits of stubble retention
Property planning	10 practices (all except trees and shrubs planted, time controlling pest plants and animals)	Seven practices (all except cropping using no-till or min-till, or gully erosion control)
Landcare participation	11 practices (all except tested water quality of main water source)	Seven practices (all except tested water quality, establishing off-stream watering, farm forestry)
Commodity group participation	10 practices (all except native vegetation fenced, time controlling pest plants and animals)	Seven practices (all except fencing wetlands, gully erosion control, establishing off-stream watering)

What we cannot change but need to consider

We need to acknowledge that there are factors beyond the control of NRM practitioners that affect landholder implementation and these need to be considered when developing policy approaches and particularly when engaging individual landholders. For example, the values that landholders attach to particular assets on their property or the district are powerful influences on behaviour (Table 11.8), and yet these values remain relatively stable. There are some consistent trends here in that respondents who attached strong environmental values to their property are significantly more likely to implement conservation practices. For example, Wimmera respondents who gave a high rating to the value 'native vegetation on my property provides habitat for native animals' were significantly more likely to fence rivers/streams/ wetlands to manage stock access and plant trees and shrubs. This was also the case for production values and CRP with a production focus. For example, Wimmera respondents who gave a high rating to the value 'the property provides most of the household income' were significantly more likely to crop using both minimum and no-till practices, sow perennial pasture and test the quality of the main water source for stock or irrigation purposes. However, there were some values that cut across the conservation-production divide. For example, those who

Table 11.8.   Significant positive relationships between topics NRM practitioners have limited capacity to influence but should consider, and implementation of CRP, Corangamite 2006 (n=482) and Wimmera 2007 (n=503).

Topics	Corangamite n=482 (12 CRP)	Wimmera n=503 (10 CRP)
Direct government support for work	All 12 practices	Eight practices (all except min-till and testing water quality),
Occupation (farmer compared to others)	10 practices (all except farm forestry, fencing native vegetation)	Six practices (all except establishing off-stream watering, gully erosion control, farm forestry, fencing wetlands)
Property profitability (reporting a profit and level of profit)	Nine practices (all except farm forestry, fencing native vegetation, time controlling pest weeds and animals)	Six practices (all except trees and shrubs planted, fencing native vegetation, gully erosion control, testing water quality)
Property size (larger)	11 practices (all except time controlling pest weeds and animals)	Eight practices (all except farm forestry, gully erosion control,)
Values attached to property	Nine of 11 practices (all except trees and shrubs planted) across 15 of 16 landholder values items	Seven practices (all except establishing off-stream watering, gully erosion control, farm forestry) across 16 of 18 items

gave a high rating to the value 'being able to pass the property on in better condition' were significantly more likely to implement almost all practices.

Occupational identity

In both the Corangamite and Wimmera studies, there were significant differences between landholders who identified as farmers by occupation and those who stated they were not a farmer. These differences included respondent concerns about issues, values, knowledge of NRM, attitudes about NRM roles and responsibilities, confidence in CRP, sources of information, property size, on-property work commitment, absentee ownership and implementation of practices (Tables 11.9 and 11.10) (Curtis *et al.* 2006; Curtis *et al.* 2008b). The narrative emerging from the analysis of survey data is consistent, if a little counterintuitive. It was not surprising to find that farmers gave higher ratings than non-farmers to the production values of their property, were more concerned about issues affecting production and were more likely to implement production related CRP. Similarly, one might have expected to find that

Table 11.9.   Differences between farmers and non-farmers: Wimmera 2007 (n=484).

Topics	Farmers	Non-farmers
Property size	880 ha	270 ha
Absentee residence	8%	50%
On-property work	50 hrs/wk	10 hrs/wk
Concern about NRM issues	Different on 14 of 21 items	
Attitudes to NRM roles and responsibilities	Different on eight of 11 items	
Knowledge of NRM	Different on 11 of 18 items	
Values attached to property	Different on 14 of 18 items	

Table 11.10.   Differences between farmers and non-farmers: Corangamite 2006 (n=453).

Topics	Farmers	Non-farmers
Property size	286 ha	40 ha
Absentee residence	12%	34%
On-property work	60 hrs/wk	10 hrs/wk
Concern about NRM issues	Different on nine of 25 items	
Attitudes to NRM roles and responsibilities	Different on 10 of 17 items	
Knowledge of NRM	Different on 14 of 17 items	
Values attached to property	Different on 16 of 21 items	

non-farmers were more concerned than farmers about conservation issues and gave a higher rating to conservation values. However, going against these intuitive expectations was the finding that farming was linked to higher implementation of some conservation practices (Table 11.8). Part of the explanation for this counter-intuitive finding is that farmers are more knowledgeable about NRM, spend more time on their property (Tables 11.9 and 11.10) and are more connected to the local community and the networks that facilitate the exchange of information and the establishment of local norms about what a 'good farm looks like in this district'. Occupation identity is also important because many non-farmers are newcomers to the area and are unable to benefit as much from existing networks.

Non-farmers represent a substantial and increasing proportion of landholders and there are significant spatial concentrations of non-farmers in particular areas. For example, 47% of Corangamite and 33% of Wimmera respondents identified as non-farmers and in the Wimmera region, the proportion identifying as a farmer declined significantly from 80% of all respondents in 2002. The proportion of non-farmers in the Wimmera study varied from 9% in the Grampians to 85% in the West Wimmera Plains and in Corangamite from 13% in Otway Coast to 88% in Lismore.

Government funding

As explained earlier, the severe drought in the Wimmera region between 2002 and 2007 contributed to significantly reduced levels of on-property profitability and median incomes. Higher levels of profitability are linked to significantly higher implementation of most CRP included in the Wimmera survey (Table 11.8). Given the relationships between income and implementation, and evidence of a decline in the proportion of landholders implementing most CRP, and the median amount of work they have completed between 2002 and 2007 in the Wimmera (most of which was not funded by government), there appears to be a strong case for increases in government funding for work with a high public benefit, including raising the cost share contributed by government during drought. While a range of delivery mechanisms might be used, a substantial part of this increased funding should be delivered through existing, effective platforms, including landcare and commodity groups.

NRM practitioners will be engaging a very different cohort of rural landholders
Rural property turnover

In this section, we describe the emerging trend of increased rural property turnover and explain its link to the influx of a different cohort of rural landholders. Our research suggests that the reconfiguration of rural space identified by others (Barr 2003; Holmes 2002, 2006), which is expected to lead to more hetereogenous 'social landscapes' (Barr 2003) and 'occupance modes', is occurring faster and more widely than anticipated. It seems that rural Australia is indeed moving towards multifunctional landscapes where a mix of production, consumption and conservation values shape land use and the landscape (Argent 2002; Cocklin *et al.* 2006; Holmes 2006; Smailes 2002). Our modelling of property turnover across a number of regions, including Corangamite and the Wimmera, suggests up to 50% of rural properties will change ownership in the next decade, that the new owners will be substantially different to the longer-term owners, and that these differences will impact on their land management. It is unlikely that 'business as usual' approaches will be effective in engaging this new landholder cohort.

A thorough explanation of the modelling approach adopted to predict turnover in property ownership is provided in Mendham and Curtis (2010) and our various technical reports (e.g. Curtis *et al.* 2008b). Since 1998, the lead author (Curtis *et al.* 2000) has been using landholder survey data (respondent's long-term plans for their property and their age) and ABS Life Expectancy Tables to model rural property turnover. An important element of the modelling is the assumption that most farmers will retire by 65 years of age. This is justified by the observation that in 2006 only 13.9% of people employed in agriculture, fishing and forestry are over 65 years (Australian Bureau of Statistics 2006).

It seems that in recent decades most of rural Victoria has had relatively stable populations. For example, our survey data show that in both the Wimmera and Corangamite regions, the median length of property ownership is close to 35 years, with very few new owners (owned property for <10 years: 15% in Wimmera; 19% in Corangamite) (Curtis *et al.* 2006, 2008b). These findings are supported by the second author's analysis of Victorian State Government property sales data for the two regions from 1995 to 2005 (Mendham and Curtis 2010). In the past 10 years, approximately 21% of properties in the Wimmera sold, while 25% sold in Corangamite. Turnover rates per year varied from 1.5% to 3.5% and the rate of turnover was trending up over time. With a median age of 54 years for Wimmera and 55 years for Corangamite respondents, it is not surprising that our modelling suggests that 50% of properties in Corangamite and 45% in the Wimmera are likely to change hands in the next 10 years (Mendham and Curtis 2010).

Newer and longer-term owners are different

Should this high level of ownership turnover occur, it is likely to lead to profound changes in the landscape (social and physical) of both regions because the new owners are typically very different from the longer-term owners. A key difference is that many of the new owners previously lived outside the district where their rural property is located. For example, analysis of Victorian property sales data indicated that 61% of new owners of rural property over the past 10 years in Corangamite and 42% in the Wimmera lived outside the local government areas where the purchased property was located. Analysis of the landholder survey data suggested that a majority of new property owners (67% in the Corangamite and 76% in the Wimmera) lived in a different district prior to purchasing their rural property.

Newer and longer-term owners were also significantly different on a range of other social and farming variables, including values attached to their property, attitudes about roles and responsibilities of NRM stakeholders, concerns about issues, confidence in CRP, knowledge of NRM and sources of NRM information (Table 11.11) (Curtis *et al.* 2006, 2008b). Many of these differences appear to be attributable to differences between respondents with farming and non-farming occupations and the fact that most new owners identified themselves as non-farmers (Table 11.11). As explained previously, farmers and non-farmers are significantly different in terms of their concern about NRM issues; their values, knowledge attitudes and sources of information; confidence in CRP and implementation of practices. Consistent with their non-farming occupations, newer owners were more likely to own smaller properties, spend less time working on-property, more time working off-property, and were less likely to make an on-property profit (Tables 11.11 and 11.12). Newer owners were also less likely to be a member of landcare or other commodity and industry groups (Table 11.11). An important finding was that a significantly larger proportion of newer owners listed their principal place of residence as being off-property.

While the dominant land use in both regions was dryland pasture, there were some significant differences between on-property enterprises of newer and longer-term landholders. In the Wimmera, longer-term landholders were more likely to be involved in cropping (78% for longer-term compared to 46%; p<0.001) and sheep production for wool (68% compared to 42%; p<0.001) or meat (70% compared to 52%; p=0.004). Newer owners were significantly more likely to be involved in alternative forms of livestock (16% compared to 7%; p=0.016). Similar trends were observed in the Corangamite region, where longer-term owners were likely to be involved in broadacre farming (29% compared to 16%; p=0.027), sheep (47% compared to 28%; p=0.003) and dairy (24% compared to 11%; p=0.01). Newer owners in Corangamite were more involved (as in the Wimmera) with other livestock (14% compared to 7%; p=0.036) as well as viticulture (12% compared to 5%; p=0.029).

The information sources used by new and longer-term owners are also different. As explained, newer owners were less likely to be members of traditional NRM networks (Table 11.11). Instead, newer owners appeared to be using more contemporary sources of

Table 11.11.   Comparing newer (<10 years) and longer-term (>10 years) landholders: social and farming variables, Corangamite 2006 (n=482) and Wimmera 2007 (n=503).

	Wimmera		Corangamite	
Topic	New (n=70)	Longer-term (n=408)	New (n=92)	Longer-term (n=381)
Median area of land managed	145 ha	722 ha***	44 ha	160 ha***
Occupation (% farmers/ non-farmers)	35%	73%***	23%	61%***
Property principal place of residence	58%	81%***	61%	81%***
Median age	48 yrs	55 yrs***	47 yrs	57 yrs***
Median hours per week on-property work	20 hrs	50 hrs***	16 hrs	40 hrs***
Median days off-property work per year	50 days	0 days**	200 days	0 days***
Made an on-property profit	17%	38%**	35%	68%***
Median level of equity in property	61–80%	81–100%*	61–80%	81–100%***
Member of landcare	17%	43%***	24%	37%*
Completed a short course related to property management	37%	50%	33%	38%
Commodity group membership	8%	27%**	13%	20%
Employed a consultant	31%	35%	29%	21%
Property management plan	early stages	early stages	not started	not started
Family members interested in taking on the property	49%	61%	36%	53%**
Succession plan	early stages	early stages	not started	early stages*

Significance denoted by: *p<0.05; **p<0.01; ***p<0.001

Table 11.12. Sources of information for newer (<10 years) and longer-term (>10 years) landholders, Corangamite 2006 (n=482) and Wimmera 2007 (n=503) (percentage of landholders using each information source).

Information sources	Wimmera (n=503)		Corangamite (n=461)	
	New (%)	Longer-term (%)	New (%)	Longer-term (%)
Television	41	45	34	33
Books/magazines/journals	65	77	73	63
CMA	43	55	33	48
Children	5	15	8	11
Victorian Farmers' Federation (VFF)	11	33***	11	32***
Bureau of Meteorology	29	31	30	30
Local council	41	45	24	22
Radio	54	61	41	47
Landcare group	40	57*	39	54*
Extension officers	30	27	14	26*
Department of Primary Industries (DPI)	41	42	24	30
Waterwatch	5	18*	3	9
Newspapers	65	83**	70	78
Field days	44	56	26	36
Environmental organisations	32	26	29	24
Industry groups	13	16	10	13
Friends/neighbours/relatives	59	54	60	50
Department of Sustainability and Environment (DSE)	43	46	37	35
Landcare coordinator	26	36	29	36
Agricultural consultants	24	22	16	15
Universities	5	5	14	5**
Internet and email	29	19	43	18***
Workshops/seminars	13	23	16	19
Mailed brochures/leaflets/ community newsletters	70	69	57	61
Training courses	8	14	17	8*

Significance denoted by: *p<0.05; **p<0.01; ***p<0.001

information than longer-term owners, including the internet (although both groups mostly used newspapers, books and mailed brochures) (Table 11.12).

Newer owners were more likely to value their properties as a place for recreation, as a break from their normal occupation, and for providing habitat for native animals. On the other hand, longer-term owners were more likely to value their properties for the social and economic outcomes linked to farming such as for providing most of the household income, providing a sense of accomplishment from building or maintaining a viable business and being able to employ family members.

There was a clear split between the two groups concerning their assessment of the importance of issues affecting their local district, with newer landholders in the Wimmera expressing significantly greater concern for nine of the ten environmental issues. In the Corangamite survey, newer landholders were significantly more concerned about four of the five environmental issues affecting their property and five of the eight environmental issues affecting the district. On the other hand, longer-term owners expressed greater concern about lower returns limiting their capacity to invest on-property.

In the Wimmera and Corangamite surveys, newer landholders were significantly more likely to agree with all three statements supporting the introduction of a landholder duty of care for the environment that was likely to impinge on private property rights (for example, 'it is fair that the wider community asks landholders to manage their land in ways that do not cause foreseeable harm to the environment' and 'in future, landholders should expect to be legally responsible for managing their land in ways that do not cause foreseeable harm to the environment'). Nevertheless, there was majority support from both groups for the proposition that 'new owners should abide by agreements entered into by previous owners where public funds have been paid for land protection or conservation work'.

Longer-term owners had significantly higher self-assessed knowledge of many of the NRM topics included in both regional surveys (12/18 items in the Corangamite survey; 7/17 items in the Wimmera survey). The small number of items where newer owners rated their knowledge higher than longer-term owners invariably related to habitat conservation. Consistent with their stronger conservation orientation, newer owners expressed significantly more interest in conservation covenants (Wimmera: new 17%, longer-term 8%; p=0.020; Corangamite: new 14%, longer-term 7%; p=0.033). Compared to longer-term owners, newer owners in the Wimmera were significantly more likely to say they would be willing to undertake environmental works on their properties without any external financial support (64% compared to 48%; p=0.020).

Analysis of survey data highlighted a trend for longer-term owners to implement CRP at higher levels (statistically significant for planting native vegetation, establishing perennial pastures and applying minimum tillage cropping techniques). These differences extended across CRP linked to both sustainable agriculture and biodiversity conservation. As with the farmer/non-farmer dichotomy, it seems that the more knowledgeable longer-term owners, who spend more time on their property and are better connected to local information and norm establishing networks, are more likely to implement most CRP.

There were also differences between new and longer-term owners in their levels of interest in a range of options for engaging landholders in NRM. In Corangamite, newer owners were more likely than longer-term owners to express interest in annual payments (37% compared to 27%; p=0.030) or a grant scheme administered by a government department (34% compared to 22%; p=0.004). In the Wimmera, newer owners were more interested than longer-term owners in training that would help them identify (35% compared to 14%; p<0.001) and establish and manage (28% compared to 17%; p<0.001) native vegetation. This interest in training is an important finding given that newer owners generally said they had less knowledge than longer-term owners. The implications of these and other findings are explored further in the conclusions section that follows.

Conclusions

Our aim in this chapter was to assist regional NRM practitioners engage rural landholders, develop a suite of policy instruments and evaluate NRM program outcomes. In concluding, we summarise our key findings and reflect on some of their implications for regional NRM practitioners.

Survey findings suggest that the voluntary contributions of landholders are critical to the success of regional NRM programs. NRM practitioners need to at least ensure they don't undermine these efforts and, if possible, they should seek to nurture them. For example, most of those implementing each CRP in the Corangamite and Wimmera studies had not received financial or technical support from government for that work. Comparison of the 2002 and 2007 Wimmera survey data showed declines in both the proportion of landholders implementing work related to most CRP and the median amount of work undertaken to enhance the condition of priority assets. It seems that this decline in landholder implementation largely resulted from a decline in voluntary efforts linked to drought and declining on-property incomes. At the same time, there was a significant positive relationship between government support and implementation. These findings support calls to increase the public share of the cost of implementing work with a high public benefit, especially during droughts.

While a range of delivery mechanisms might be used, a substantial part of this increased funding should be delivered through existing, effective platforms, including landcare and commodity groups. Again, research presented here provides considerable evidence of the positive impact of regional NRM investments in capacity building focused on raising awareness, improving knowledge and building confidence in CRP – including through short courses such as those undertaken as part of property management planning.

There are also factors beyond the control of NRM practitioners that affect landholder implementation. Landholder values are powerful influences on behaviour and yet remain relatively stable. Accordingly, NRM practitioners need to be aware of the range of landholder values and ensure that their communication and policy approaches embrace those values. While there is an apparent division between conservation and production values, some values cut across the conservation-production divide. For example, those who gave a high rating to the value 'being able to pass the property on in better condition' were significantly more likely to implement almost all CRP included in the surveys.

In both the Corangamite and Wimmera studies, there were significant differences between landholders who identify as farmers by occupation and those who say they are not a farmer. Farmers consistently gave higher ratings to the production values of their property, were more concerned about issues affecting production and were more likely to implement production related CRP. Compared to farmers, non-farmers were more concerned about conservation issues and gave a higher rating to conservation values. However, there was higher implementation of some conservation practices amongst farmers than non-farmers. Part of the explanation for this counter-intuitive finding is that farmers are more knowledgeable of NRM, spend more time on their property and are more connected to the local community and the networks that facilitate the exchange of information and the establishment of local norms about what a 'good farm looks like in this district'.

Evidence presented suggests that we can expect a significantly higher rate of change in ownership of rural properties in Victoria over the next decade. This is likely to have profound implications for communities, industries and NRM. The scale and nature of predicted property ownership change can be expected to reduce the effectiveness of local groups and their networks if they lose key leaders or they don't have sufficient numbers of people living and working full-time in the local community to share the burden of community work. These changes are likely to impact on the capacity of landcare groups to function and implement on-ground work, but also on the viability of local fire brigades and service groups that are critical elements of rural life.

The new owners are typically very different from the longer-term owners in that most of them are not farmers by occupation, are more likely to have resided outside the district prior to purchasing their rural property and to be absentee owners. Newer and longer-term owners were also significantly different in terms of the values attached to their property, attitudes about roles and responsibilities of NRM stakeholders, concerns about issues, confidence in CRP, and sources of NRM information. While newer owners can be expected to bring new ideas, skills and access to resources, they are less knowledgeable about NRM and have less experience of local conditions, including droughts.

Our contact with NRM practitioners indicates they are experiencing difficulties engaging many newcomers in NRM. A 'business as usual' approach involving appeals and policy instruments designed for full-time farmers, largely motivated by a desire to increase production and profitability, is unlikely to motivate the increasingly important cohort of newcomers. Appeals that address the newcomers' pro-conservation values are more likely to be successful. New landholders also expressed a specific interest in conservation covenants and in training that would address their need for more knowledge on how to identify, establish and manage native vegetation. If NRM agencies have limited resources for native vegetation management, they might therefore be advised to invest in training programs rather than covering the cost of on-ground work for a relatively affluent cohort of landholders, most of whom say they would undertake conservation work without government funding.

The high proportion of new owners who are absentee residents also presents a challenge for NRM practitioners. Many absentees are unable to participate in typical evening meetings or the daytime workshops and field days held as part of traditional group or agency extension programs. An obvious strategy is to hold these activities on weekends and to promote activities and access to information using a variety of media, including web-based formats. Even then, absentees may be reluctant to invest time in meetings or workshops unless they address their specific needs. NRM practitioners need to be active in identifying newcomers, perhaps at the time of a property sale, and then make personal contact to identify their needs and provide them with ways to engage local networks and wider information sources.

References

ANAO (Australian National Audit Office) (1997) 'Commonwealth natural resource management and environment programs'. Audit Report No. 36 1996–97. Australian National Audit Office: Canberra.

ANAO (Australian National Audit Office) (1998) 'Preliminary inquiries into the National Heritage Trust'. Audit Report No. 42 1997–98. Australian National Audit Office: Canberra.

ANAO (Australian National Audit Office) (2001) 'Performance information on Commonwealth financial assistance under the Natural Heritage Trust'. Audit Report No.43 2000–01. Australian National Audit Office: Canberra.

ANAO (Australian National Audit Office) (2008) 'Regional delivery model for the National Heritage Trust and the National Plan for Salinity and Water Quality'. Audit Report No. 21 2007–08. Australian National Audit Office: Canberra.

Argent N (2002) From pillar to post? In search of the post-productivist countryside in Australia. *Australian Geographer* **33**(1), 97–114.

Australian Bureau of Statistics (2002) 'Measures of Australia's progress 2002 (Catalogue No. 1370.0)'. ABS: Canberra.

Australian Bureau of Statistics (2006) 'Basic community profile Victoria: industry of employment by age and sex'. Australian Bureau of Statistics: Canberra.

Bammer G, Mobbs C, Lane R, Dovers S and Curtis A (2005) An introduction to Australian case studies of integration in natural resource management. *Australasian Journal of Environmental Management* **12** (supplementary issue), 3–5.

Barr N (2003) Future agricultural landscapes. *Australian Planner* **40**(2), 123–127.

Barr N and Cary J (1992) *Greening a Brown Land: The Australian Search for Sustainable Land Use*. Macmillan: South Melbourne.

Barr N and Cary J (2000) *Influencing Improved Natural Resource Management on Farms*. Bureau of Rural Sciences: Canberra.

Barr N, Ridges S, Anderson N, Gray I, Crockett J, Watson B and Hall N (2000) *Adjusting for Catchment Management: Structural Adjustment and Its Implications for Catchment Management in the Murray Darling Basin*. Murray-Darling Basin Commission: Canberra.

Black A and Hughes P (2001) *The Identification and Analysis of Indicators of Community Strength and Outcomes*. Department of Family and Community Services: Canberra.

Byron I and Curtis A (2002a) *Landcare Participation and Outcomes in the Lachlan Catchment*. Johnstone Centre, Charles Sturt University: Albury, NSW.

Byron I and Curtis A (2002b) *Landcare Participation and Outcomes in the Queensland Murray-Darling Basin*. Johnstone Centre, Charles Sturt University: Albury, NSW.

Cary J, Webb T and Barr N (2002) *Understanding Landholders' Capacity to Change to Sustainable Practices: Insights About Practice Adoption and Social Capacity for Change*. Bureau of Rural Sciences: Canberra.

Castle EN (2002) Social capital: an interdisciplinary concept. *Rural Sociology* **67**(3), 331–349.

Cavaye J (2003) *Integrating Economic and Social Issues in Regional Natural Resource Management Planning: A Framework for Regional Bodies (National Action Plan for Salinity and Water Quality)*. Queensland Department of State Development: Brisbane.

Cocklin C, Dibden J and Mautner N (2006) From market to multifunctionality? Land stewardship in Australia. *The Geographical Journal* **172**(3), 197–205.

Cook TD and Shadish WR (1986) Program evaluation: the worldly science. *Annual Review of Psychology* **37**, 193–232.

Corangamite Catchment Management Authority (2003) 'Corangamite Regional Catchment Strategy 2003–2008'. Corangamite Catchment Management Authority: Colac, Victoria.

Curtis A (2007) Monitoring and evaluation of watershed initiatives: lessons from Landcare in Australia. In *Monitoring and Evaluation of Soil Conservation and Watershed Development Projects*. (Eds J de Graaff, J Cameron, S Sombatpanit, C Pieri and J Woodhill) pp. 377–397. Science Publishers: New Hampshire, USA.

Curtis A and Byron I (2002) *Understanding the Social Drivers of Catchment Management in the Wimmera Region*. Johnstone Centre, Charles Sturt University: Albury, NSW.

Curtis A and De Lacy T (1996) Landcare in Australia: does it make a difference? *Journal of Environmental Management* **46**(2), 119–137.

Curtis A and Lockwood M (2000) Landcare and catchment management in Australia: lessons for state-sponsored community participation. *Society and Natural Resources* **13**(1), 61–73.

Curtis A and Robertson A (2003a) Understanding landholder management of river frontages: the Goulburn Broken. *Ecological Management and Restoration* **4**(1), 45–54.

Curtis A and Robertson A (2003b) Who are program managers working with and does it matter? The experience with river frontage management in the Goulburn Broken. *Natural Resource Management* **6**(2), 25–32.

Curtis A, Byron I and MacKay J (2005) Integrating socio-economic and biophysical data to underpin collaborative watershed management. *Journal of the American Water Resources Association* **41**(3), 549–563.

Curtis A, Lockwood M and MacKay J (2001) Exploring landholder willingness and capacity to manage dryland salinity in the Goulburn-Broken Catchment. *Australian Journal of Environmental Management* **8**(2), 79–90.

Curtis A, Race D and Robertson A (1998) Lesson from recent evaluations of natural resource management programs in Australia. *Australian Journal of Environmental Management* **5**(2), 109–119.

Curtis A, MacKay J, Van Nouhuys M, Lockwood M, Byron I and Graham M (2000) *Exploring Landholder Willingness and Capacity to Manage Dryland Salinity: The Goulburn Broken Catchment*. Johnstone Centre, Charles Sturt University: Albury, NSW.

Curtis A, Cooke P, McDonald S and Mendham E (2006) 'Corangamite region social benchmarking survey 2006'. Institute for Land, Water and Society, Charles Sturt University: Albury, NSW.

Curtis A, Lucas D, Nurse M and Skeen M (2008a) *Achieving NRM Outcomes through Voluntary Action: Lessons from Landcare*. Department of Sustainability and Environment: Melbourne.

Curtis A, McDonald S, Mendham E and Sample R (2008b) *Understanding the Social Drivers for Natural Resource Management in the Wimmera Region*. Institute for Land, Water and Society, Charles Sturt University: Albury, NSW.

Dillman DA (1978) *Mail and Telephone Surveys: The Total Design Method*. John Wiley: New York.

Dovers SR and Mobbs CD (1997) An alluring prospect? Ecology, and the requirements of adaptive management. In *Frontiers in Ecology: Building the Links*. (Eds N Klomp and ID Lunt) pp. 39–52. Elsevier: Oxford.

Guerin LJ and Guerin TF (1994) Constraints to the adoption of innovations in agricultural research and environmental management: a review. *Australian Journal of Experimental Agriculture* **34**(4), 549–572.

Holmes J (2002) Diversity and change in Australia's rangelands: a post-productivist transition with a difference? *Transactions of the Institute of British Geographers* **27**(3), 362–384.

Holmes J (2006) Impulses towards a multifunctional transition in rural Australia: gaps in the research agenda. *Journal of Rural Studies* **22**(2), 142–160.

Mendham E and Curtis A (2010) Taking over the reins: trends and impacts of changes in rural property ownership. *Society and Natural Resources* **23**(7), 653–668.

Mendham E, Millar J and Curtis A (2007) Landholder participation in native vegetation management in irrigation areas. *Ecological Management and Restoration* **8**(1), 42–48.

Millar J and Curtis A (1997) Perennial grasses: finding the balance. *Australian Journal of Soil and Water Conservation* **10**(1), 21–28.

Pannell DJ, Marshall GR, Barr N, Curtis A, Vanclay F and Wilkinson R (2006) Understanding and promoting adoption of conservation technologies by rural landholders. *Australian Journal of Experimental Agriculture* **46**(11), 1407–1424.

Paton S, Curtis A, McDonald GT and Woods M (2004) Regional natural resource management: is it sustainable? *Australian Journal of Environmental Management* **11**(4), 259–267.

Patton MQ (1990) *Qualitative Evaluation and Research Methods*. 2nd edn. Sage: Newbury Park, California.

Race D and Curtis A (1998) Socio-economic considerations for regional farm forestry development. *Australian Forestry* **60**(4), 233–239.

Shultz S, Saenz F and Hyman G (1998) Linking people to watershed protection planning with a GIS: a case study of a central American watershed. *Society and Natural Resources* **11**(7), 663–675.

Smailes PJ (2002) From rural dilution to multifunctional countryside: some pointers to the future from South Australia. *Australian Geographer* **33**(1), 79–95.

Sobels J, Curtis A and Lockie S (2001) The role of Landcare group networks in rural Australia: exploring the contribution of social capital. *Journal of Rural Studies* **17**(3), 265–276.

Vanclay F (1992) Barriers to adoption: a general overview of the issues. *Rural Society* **2**(2), 10–12.

Vanclay F (2004) Social principles for agricultural extension to assist in the promotion of natural resource management. *Australian Journal of Experimental Agriculture* **44**(3), 213–222.

Webb TJ and Curtis A (2002) *Mapping Regional Capacity*. Bureau of Rural Sciences: Canberra.

12

Policy perspectives on changing land management

David J Pannell

Summary

There are various policy mechanisms that may be used to encourage changes in land management by rural landholders. The choice of policy mechanism for particular programs and projects needs to be considered on a case-by-case basis. In particular, it should depend on the levels of public and private benefits that would be generated by the proposed changes in land management. The private benefits of adopting new land management practices (or in other words the adoptability of those practices) is often not adequately considered by policy makers or natural resource management bodies when selecting policy mechanisms for particular projects or programs.

Introduction

Many government programs around the world have been created to attempt to encourage landholders to change their practices in various ways. Examples with an environmental emphasis include the Conservation Reserve Program, the Environmental Quality Incentive Reserve Program, and the Conservation Security Program in the United States of America; the Rural Development Regulation in the European Union; the National Farm Stewardship Program in Canada; and the Natural Heritage Trust in Australia.

Questions that the designers and implementers of these programs need to address include:

- How should the resources of the program be targeted? Should they be directed to landholders in particular locations, or particular types of farms?
- Which policy mechanisms should be used to encourage practice change?

This chapter explores these questions. It starts by considering the justification for governments seeking rural practice change in general, followed by a discussion of issues related to the targeting of resources. Then it presents a framework for selecting policy mechanisms depending on the levels of public and private benefits that are expected to arise from the proposed changes in practices.

Justifications for governments seeking practice change

Why should governments intervene to influence landholder practices? In principle, why should landholders not be free to manage their land as they see fit? Economists recognise two broad categories of potential reasons: (a) to increase the overall benefits to the community, and (b) to redistribute benefits and costs within the community in socially desirable ways.

If markets are complete and working well, there is unlikely to be an opportunity to increase overall benefits through government intervention. It would be preferable to leave decisions about which practices to use up to individual landholders, who would be best placed to judge which practices would be in their own interests.

In practice, there are a number of reasons why markets may fail to deliver this ideal outcome, which economists discuss under the label 'market failure'. Market failure is defined as a situation where people acting independently and individually will not result in the greatest possible benefits for society, at least if the aim is to maximise the overall benefits rather than to redistribute them. If the aim is to maximise the overall benefits of actions, economists see market failure as being necessary for government action to be justified. For example, they want to see evidence of market failure before they would sanction the use of a regulation or an economic incentive payment to try to change people's behaviour in a particular situation. Without market failure, the argument goes, government action can only make things worse, because a free market would result in the best possible set of outcomes.

The idea is an important one as it provides a safeguard against wasting public money. It does this by forcing each issue to be looked at from an overall perspective, not just from a narrow sectional perspective. Without the discipline provided by rules like this, the wasteful use of public funds would be that much greater.

There are various potential causes of market failure. Three key types of market failure that are relevant to land management are outlined below.

Externalities

An externality occurs when an activity undertaken by an individual has side effects on others that are not taken into consideration by the first individual (Randall 1987). There are two types of externality: negative and positive (also called external costs and external benefits). For example, suppose pollution is generated as a side effect of agriculture. In an unregulated market, a negative externality such as this pollution is a potential problem because the level of the activity chosen by the polluters may be too great from a social point of view (in the sense that there exists the potential to improve the welfare of both the polluter and the sufferer). If the external costs could be factored into the polluters' business decisions, the polluting activity would probably be undertaken at a lower level. In the absence of regulation or some form of government-imposed incentive, the group of polluters generates more pollution than is socially desirable because they do not consider the costs it imposes on others.

A positive externality is also a potential problem, but this time because the level of the activity is too low. For example, if planting trees on a farm has a side effect of lowering salinity on a neighbouring farm, the level of tree planting may be lower than would be optimal overall.

Related to this is the issue of long time lags. If there is a long delay between the implementation of management actions to conserve a resource and the delivery of benefits from those actions, it may be that the benefits and costs are borne by different generations. It is likely that distant future benefits will not be fully reflected in current market signals, resulting in under-investment in current resource conservation actions. This may be relevant to problems such as dryland salinity, which has time lags measured in centuries in some cases.

Public goods

Public goods do not necessarily relate to 'the public' as any identifiable group. They are simply goods that have particular characteristics that mean it is often worthwhile for government to intervene on behalf of the public in some way. There are two types of public goods: non-price-excludable goods and non-rival goods (Randall 1987).

A non-price-excludable good is one that consumers cannot be prevented from consuming, leading to problems of free-riders, under-provision or over-exploitation. Consumers have open access to the good and the provider or owner of the good is unable to charge a fee for access (or there is no owner). For example, information generated by agricultural research can spread throughout the farming population without the researcher being able to charge a fee to those who benefit from the information.

For a non-rival good, consumption by one person does not reduce the availability of that good to others. For example, suppose a landholder protects biodiversity on his or her land, and that this has benefits for many people in the community who care about nature. The fact that one nature lover benefits from the landholder's actions does not reduce the benefits to other nature lovers; the good is not used up in the course of generating its benefits, in the same way as a physical good like a food product would be. Therefore, there is no cost of providing a non-rival good to an additional consumer. Economic theory says that if providing a good to an extra consumer costs nothing, then the price charged to consumers should be zero, and that is the problem. If we enforce a zero price, the supplier of the good (e.g. the landholder in this case) will not have the incentive to supply an appropriate amount of the good.

Information asymmetry

The government may have information about a land management issue that is not known to at least some landholders. For example, research may have revealed that a new type of crop that is not yet widely adopted is highly beneficial to landholders (as occurred with lupins in Western Australia in the 1980s, Marsh *et al.* 2000). There would be benefits from government taking action to promote its rapid adoption. The information asymmetry may also operate in the other direction. In considering whether a particular type of practice change should be promoted, landholders are likely to have better information about the costs to them of making the change than does government. This is not so much a cause of market failure as a constraint on the design of appropriate government policy. One effective response has been to use market-like mechanisms, such as conservation tenders, to reveal this hidden information (e.g. Stoneham *et al.* 2003).

Targeting policy interventions

When deciding how government funds should be targeted, the market-failure criterion is important, but it is a not a sufficient guide to action. For example, there are many situations where land management causes externalities and public-good problems due to environmental impacts, but not all of them are equally deserving of government attention. Where program budgets are limited, a more sophisticated approach to targeting investment, involving a more detailed assessment of priorities, is required. Basically, the assessment should consider the benefits and costs of alternative projects, and select those projects with the highest net benefits per dollar of public expenditure. Ideally, Benefit:Cost Analysis (e.g. Brouwer and Pearce 2005; Mishan and Quah 2007) would be conducted as part of each project's assessment. In practice, this rarely happens.

Various other assessment methods have been developed. For example, Pannell *et al.* (2009) developed the Investment Framework for Environmental Resources (INFFER) as a simplified tool that is consistent with Benefit:Cost Analysis. The criteria they use to estimate benefits and costs of environmental projects include:

- relative significance or value of the relevant environmental assets
- severity of existing or threatened damage to the assets
- technical feasibility of reversing or preventing that damage
- urgency of taking action
- negative spin-offs from the proposed actions
- costs of required policy interventions
- adoptability of desired new practices.

The adoptability of a desired new practice is important to policy makers for two reasons: (a) it influences the cost of achieving the required level of adoption, with more adoptable practices being less costly to promote successfully; (b) it influences the appropriate choice of policy mechanism (see next section). Often programs neglect this consideration when selecting priority interventions and policy mechanisms (Pannell *et al.* 2006), resulting in disappointing levels of outcomes.

Choosing types of policy mechanisms

There is a tendency for some people to think of extension (including awareness raising, technology transfer, training, etc.) as the default policy mechanism to encourage practice change. However, it is not the only option, and in many situations it is not the most appropriate mechanism for promoting change.

In government programs, the choice among the possible policy mechanisms (Table 12.1) is often not very sophisticated. Programs tend to rely primarily on a small number of mechanisms, sometimes as few as one. Here, based on Pannell (2008, 2009), it is shown that the choice between policy mechanisms can be greatly helped by an appreciation of the levels of public and private net benefits that are likely to result. This framework is designed for programs where the aim is to address market failure due to externalities, such as environmental benefits and costs arising from land-use change (e.g. Roberts and Pannell 2009). The discussion below is couched in terms of a project that attempts to influence land management in order to increase environmental benefits or reduce environmental costs.

'Private net benefits' refer to benefits minus costs accruing to the private land manager as a result of the proposed changes in land management. 'Public net benefits' are benefits minus costs accruing to everyone other than the private land manager. Defining them in these ways is

Table 12.1. Alternative policy mechanisms for seeking changes in management of private lands.

Category	Specific policy mechanisms
Positive incentives	Financial or regulatory instruments[a] to encourage change
Negative incentives	Financial or regulatory instruments[a] to inhibit change
Extension	Technology transfer, education, communication, demonstrations, support for community network
Technology change	Development of improved land management options, such as through strategic R&D, participatory R&D with landholders, provision of infrastructure to support a new management option
Research	Provision of information relevant to the assessment of various practices
No action	Informed inaction

[a] Financial or regulatory instruments include polluter-pays mechanisms (command and control, pollution tax, offsets), beneficiary-pays mechanisms (subsidies, conservation auctions and tenders), and mechanisms that can work in either way depending on how they are implemented (define and enforce property rights, such as through tradeable permits).

helpful because the private net benefit dimension provides insight into the behaviour of the landholder, while the public net benefit dimension relates to the effects on everyone else that flow from the landholder's behaviour. Defined this way, public benefits are equivalent to external benefits (see earlier section 'Justifications for governments seeking practice change').

The private net benefits of a project (i.e. a specific set of land-use changes) depend on:

- the financial returns from the new land uses
- the financial returns from the land uses that are replaced (the 'opportunity costs')
- any change in the level of risks faced as a result of the change
- indirect impacts on other aspects of the farm system or on the farmer's lifestyle
- the farmer's own interest in environmental outcomes.

The public net benefits depend on:

- the value or importance of the environmental assets that are affected by the changes
- the degree of degradation that the assets were facing or had already suffered
- the extent to which that degradation can be prevented or alleviated by the changes
- any lags in the response of the biological or physical system to the land-use changes.

In the graphs that follow, the net benefits relate to the benefits and the costs of the proposed land-use changes, assuming that those changes occur. They exclude costs of financial transfers from the environmental manager to landholders to encourage the change in land management. This allows us to compare the benefits of an intervention with its costs.

Having established the definitions, the starting point for the framework is the recognition that environmental managers can invest in a range of projects involving changes in land management or land use on private land, and that the available options vary widely in the levels of public and private net benefits they generate, potentially including negative net benefits. Compared to the *status quo*, a given set of changes in land management could have either positive public net benefits (areas F, A and B in Figure 12.1) or negative public net benefits (areas C, D and E), and either positive private net benefits (areas B, C and D) or negative private net benefits (areas E, F and A). Combining public and private outcomes, the changed management practices may be positive overall (areas A, B and C), if, for instance, there are positive private net benefits that exceed negative public net benefits, as in area C, or negative overall (areas D, E and F).

The aim is to allocate the categories of mechanisms in Table 12.1 to the six areas in Figure 12.1. It is assumed that landholders will adopt all land-management practices with positive private net benefits (projects in areas B, C, D) provided that they are able to learn about those practices. Initially, zero learning costs for landholders are assumed. Given this assumption, a total of 10 rules for selecting policy mechanisms are proposed. Rules 1 to 3 apply to the use of positive incentives:

(1)   Do not use positive incentives for land-use change unless public net benefits of change are positive: no positive incentives for C, D, E.
(2)   Do not use positive incentives if landholders would adopt land-use changes without those incentives: no positive incentives for B.
(3)   Do not use positive incentives if private net costs outweigh public net benefits: no positive incentives for F.

These rules narrow the use of positive incentives to area A. Now consider Rules 4 and 5 for the use of extension. These rules are based on the use of extension to improve decision making, rather than to improve skills. They relate to the use of extension as the main policy tool, rather than as a support to other policy mechanisms.

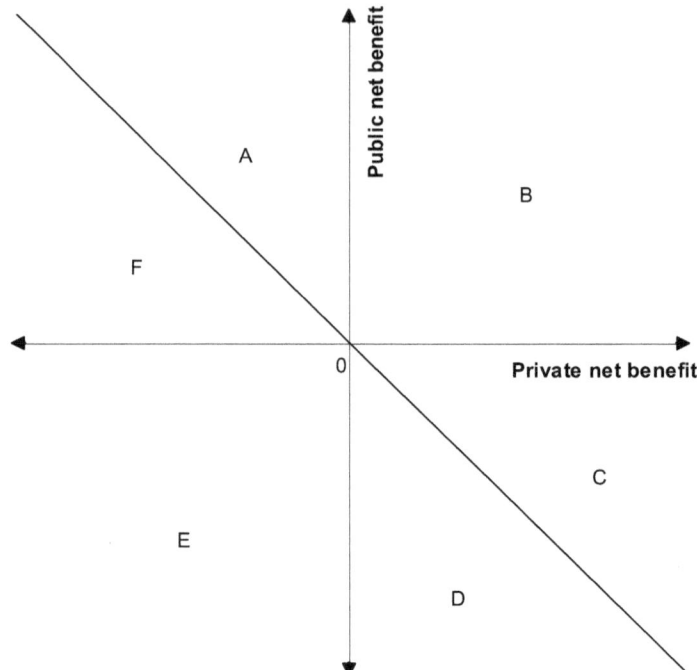

Figure 12.1. Possible combinations of public and private benefits from changing land management practices away from the status quo.

(4) Do not use extension unless the change being advocated would generate positive private net benefits. In other words, the practice should be sufficiently attractive to landholders for it to be 'adoptable' once the extension program ceases.

(5) Do not use extension where a change would generate negative net public benefits.

These rules narrow extension down to area B. Rules 6 to 10 apply to the use of negative incentives, technology change and no action:

(6) If private net benefits are negative and public net benefits are positive (areas F and A), consider technology development to create improved (environmentally beneficial) land management options that can be made adoptable (with or without positive incentives). The merits of technology change depend on the public and private benefits of the best available projects, and how much these can be improved through technology change (the dashed line in Figure 12.2 represents variability between projects in the extent to which benefits could be increased. The greater the increase in benefits, the further to the left would be the dashed line).

(7) If private net benefits outweigh public net costs (area C), then land-use changes should be accepted if they occur, implying no action. Alternatively, if it is not known whether private net benefits are sufficient to outweigh public net costs, a relatively flexible negative incentive instrument may be used to communicate the public net costs to land managers (e.g. a pollution tax), leaving the ultimate decision to the land managers. Inflexible negative incentives, such as command and control, should not be used in this case.

(8) If public net costs outweigh private net benefits (area D), use negative incentives to discourage uptake of the land management practices.

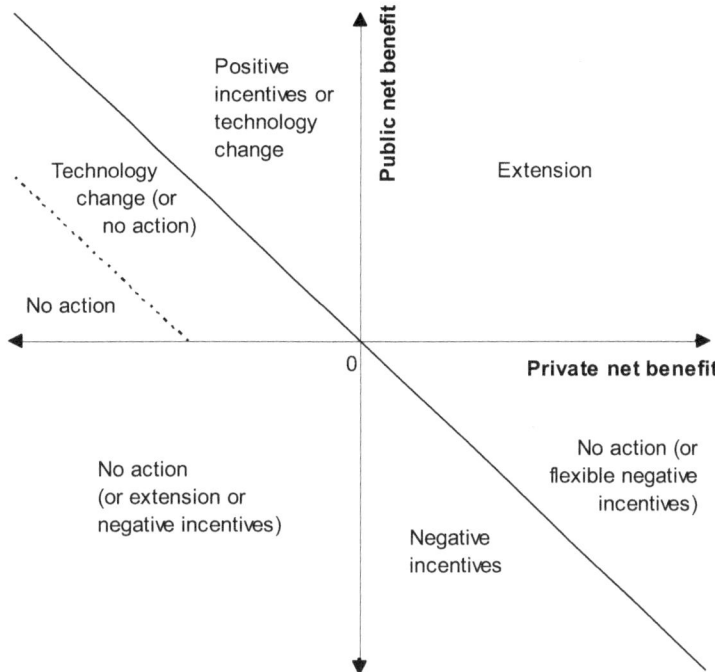

Figure 12.2. Suggested main policy mechanisms depending on public and private net benefits.

(9)   If public net benefits and private net benefits from a set of land-use changes are both negative (area E), and landholders accurately perceive this, then no action is necessary. Adverse practices are unlikely to be adopted. If there is concern that landholders have misperceptions about the relevant land uses, adoption of environmentally adverse practices could be discouraged by extension, or by negative incentives.

(10) In all cases, the suggested action needs to be weighed up against a strategy of no action.

These rules lead to Figure 12.2. Research to provide improved information is relevant to all cases, so it is not allocated to a particular area on the graph. Broadly speaking, the framework suggests the use of extension as the main policy mechanisms where public and private net benefits are both positive. Positive incentives to encourage change are suggested where public net benefits are positive and private net benefits are slightly negative. Negative incentives to discourage change are suggested where public net benefits are negative and private net benefits are slightly positive. Lastly, technology change is suggested where public net benefits are positive but private net benefits are negative.

The framework shown in Figure 12.2 is simplified in a number of respects. Figure 12.3 shows a version that includes additional complexities:

- Transition from an existing practice to another is not costless. It involves learning costs, and may require other costs such as purchase of new machinery.
- Adoption does not occur instantly, and the lag until adoption is likely to depend on the level of private net benefits.
- Extension does not eliminate the lag to adoption, but may reduce it.
- Implementing programs to achieve practice change involves transaction costs to the environmental manager.

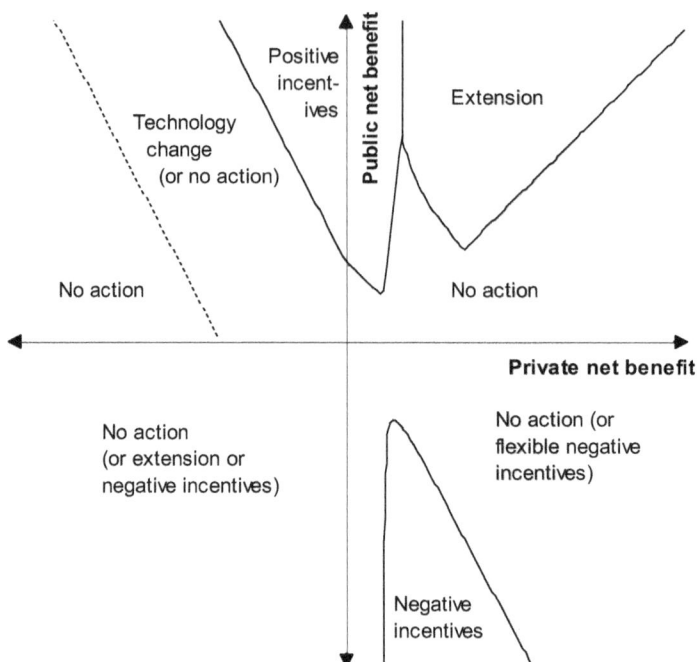

Figure 12.3. Suggested main policy mechanisms depending on public and private net benefits, accounting for additional complexities.

- Given limited budgets, environmental managers may require benefits to more than cover costs. The figure shows results for a Benefit:Cost threshold of 2:1.

Compared to Figure 12.2, Figure 12.3 is different in that: (a) public benefits need to be a little higher to justify positive incentives; (b) extension is not worthwhile if public and private net benefits are both low, nor if public net benefits are low but private net benefits are high. Extension is replaced by positive incentives if there are low private net benefits; (c) the area for negative incentives narrows to a set of projects with low private net benefits and more negative public net benefits. Broadly speaking, the two figures are similar, but intervention is more targeted in Figure 12.3.

The framework has strong implications for the targeting of positive incentives, negative incentives and extension. If they are to generate substantial net benefits, these instruments need to be carefully applied to projects that fall in the indicated areas. This presumes that there are, in fact, projects available within those areas. If not, then the use of these policy instruments is not appropriate.

It is notable that the areas for positive incentives, negative incentives and extension are small subsets of the total. A project chosen at random has only a small chance of falling into any one of these areas. Environmental managers need to take care to ensure that they are not applying these mechanisms inappropriately to projects. There seems a high risk of this unless they have very good information about both the public and private net benefits, or use a mechanism that reveals this information.

In applying the framework to case studies in Australia, it has become apparent that inappropriate policy mechanisms have often been used (Ridley and Pannell 2005). The consequence has been low levels of land-use change, and failure of the program to meet its environmental targets.

In my experience, environmental managers do try to consider the environmental (public) benefits of their funded works, but often neglect the private benefits and costs. This framework reveals that the selection of cost-effective environmental projects is perhaps even more sensitive to private net benefits than to public net benefits.

A notable implication from the framework is that projects are more likely to generate high payoffs to investment in positive incentives if the private net benefits are reasonably close to zero. If they are close to zero, land-use change can be prompted with small incentives. For extension, the most appropriate projects are those with low to moderate positive private net benefits. Projects with high private net benefits do not need extension to promote them – they will be adopted rapidly anyway. Similarly, negative incentives to prevent adoption of practices with negative public benefits are best suited to cases where the private net benefits of land-use change are positive, but only small to moderate in scale. If the private benefits are too large, it will be too difficult to prevent adoption, and it might even be undesirable overall to prevent adoption.

Pannell and Wilkinson (2009) adapted the framework to analyse policy mechanism choice for smallholder 'lifestyle' landholders. They found that, when factors such as the higher transaction costs and higher learning costs per unit area are accounted for, prospects for worthwhile public investments in land-use changes by lifestyle landholders are lower than for commercial landholders.

Adoptability

The public: private benefits framework (see previous section) highlights that the appropriate policy response depends crucially on the private net benefits of proposed new practices – in other words, the adoptability of the works being promoted by those projects. Figure 12.4 shows another way of considering policy responses based around this issue. It starts by asking whether the land management changes being proposed are adoptable on the scale required to achieve the project's goals. If they are, the next question is why they have not already been adopted. Depending on the answer to this question, different policy responses are suggested.

One possibility is that the practices are new such that the adoption diffusion process is still incomplete. Adoption would be expected to occur once landholders become fully aware of the practices. The appropriate policy response in this case is probably an extension program to raise general awareness of the practices.

If the practices are not new, but they are still not adopted despite being assessed as being adoptable, it may be that there is a learning failure. Landholders may have developed misperceptions or misconceptions about the practices. For example, Llewellyn et al. (2004) found that many farmers in a sample of Western Australian grain growers held inaccurate views about the likelihood of herbicide-resistant weeds returning to being susceptible to herbicide if the use of herbicides was suspended. They also held views about the likelihood of herbicides with new modes of action becoming available that were at odds with expert opinion. In most other respects their knowledge and perceptions about herbicide resistance and its management were reasonably accurate. Llewellyn et al. (2004) highlighted that extension would need to focus on the inaccurately perceived variables (especially those that would be decisive in shaping behaviour), rather than other variables that were accurately perceived. In other words, extension would be targeted to address specific learning failures, rather than general awareness.

The third potential reason for non-adoption of an adoptable practice is lack of skills or resources on the part of the landholders. In the case of a skills shortage, the obvious response is to provide training. In the case of lack of resources, it may be adequate to wait for a profitable season and allow landholders to adopt after that.

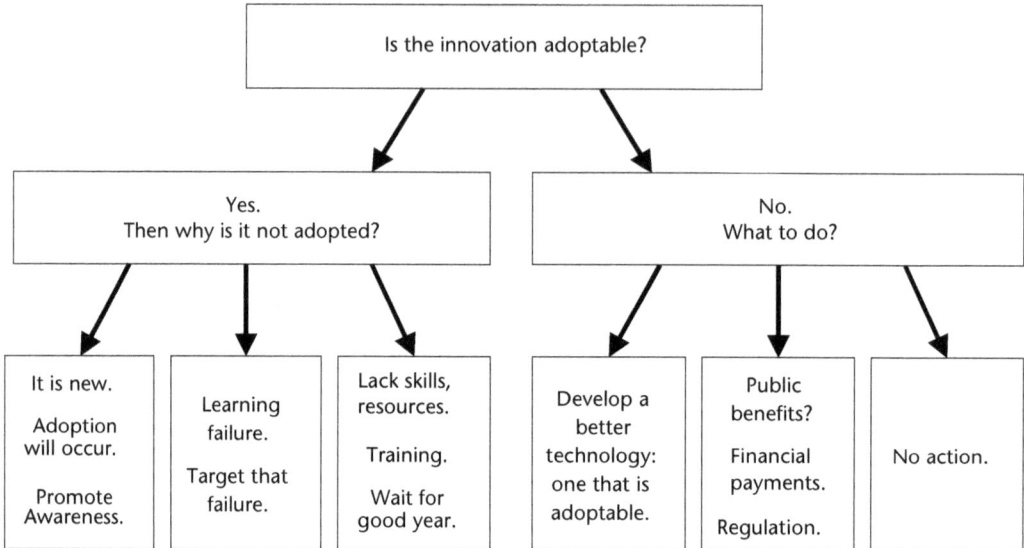

Figure 12.4. Choice of policy mechanism depends crucially on the adoptability of the proposed new practice.

If the practice is not adoptable at the required scale, one option is to attempt to develop a new practice or technology that is sufficiently adoptable. For example, if there is a desire to see farmers in a region establish perennial pastures over a large area but the available perennial pastures are assessed as being not adoptable, scientists might be funded to develop an improved type of pasture or a new pasture-production system that would be adoptable.

If the practices are not adoptable but public benefits from their adoption would be large, another option is to use an incentive-based policy mechanism to motivate their more rapid and more extensive adoption. This could include either a polluter-pays mechanism (such as a pollution tax or a command and control system) or a beneficiary-pays mechanism (such as a stewardship payment, income from a conservation auction, or an incentive payment). This is appropriate provided that the level of payment required is no so large as to outweigh the resulting environmental benefits.

Finally, there is the option of undertaking no action. Given the limited funds available in most programs, this is necessarily a common outcome.

Implications for policy

A number of policy implications have been highlighted in the previous discussion. The concept of market failure as a justification for policy intervention to encourage changes in land-management practices has been described. Economists argue that without clear evidence that a project addresses an area of market failure, net benefits from the project are unlikely.

The public: private benefits framework provides a tool to analyse appropriate categories of policy mechanisms for different projects. It indicates that projects best suited to the use of positive incentive mechanisms would be those with high public net benefits and private net benefits around zero. Negative incentives are recommended for projects with negative public net benefits, and low to moderate private net benefits. Extension is recommended as the main response for projects with positive public net benefits and moderately positive private net benefits. Technology change (primarily through research) is recommended for projects with positive public net benefits but negative private net benefits.

The great importance of private net benefits (or, in other words, 'adoptability') as a driver of the appropriate policy response has been highlighted. There are several policy responses that may be appropriate when the practices being targeted by policy are adoptable, and a different set of policy responses when they are not.

More broadly, this chapter has highlighted that the choice of policy mechanisms to encourage changes in land management need to be considered on a case-by-case basis. What works cost-effectively for one project may not work at all for another.

References

Brouwer R and Pearce D (Eds) (2005) *Cost-Benefit Analysis and Water Resources Management.* Edward Elgar: Cheltenham, UK.

Llewellyn RS, Lindner RK, Pannell DJ and Powles SB (2004) Grain grower perceptions and the use of integrated weed management. *Australian Journal of Experimental Agriculture* **44**(10), 993–1001.

Marsh SP, Pannell DJ and Lindner RK (2000) The impact of agricultural extension on adoption and diffusion of lupins as a new crop in Western Australia. *Australian Journal of Experimental Agriculture* **40**(4), 571–583.

Mishan EJ and Quah E (2007) *Cost Benefit Analysis.* 5th edn. Routledge: London.

Pannell DJ (2008) Public benefits, private benefits, and policy intervention for land-use change for environmental benefits. *Land Economics* **84**(2), 225–240.

Pannell DJ (2009) Technology change as a policy response to promote changes in land management for environmental benefits. *Agricultural Economics* **40**(1), 95–102.

Pannell DJ and Wilkinson R (2009) Policy mechanism choice for environmental management by non-commercial 'lifestyle' rural landholders. *Ecological Economics* **68**(10), 2679–2687.

Pannell DJ, Marshall GR, Barr N, Curtis A, Vanclay F and Wilkinson R (2006) Understanding and promoting adoption of conservation practices by rural landholders. *Australian Journal of Experimental Agriculture* **46**(11), 1407–1424.

Pannell DJ, Roberts AM, Alexander J and Park G (2009) 'INFFER (Investment Framework For Environmental Resources), INFFER Working Paper 0901'. University of Western Australia: Perth. http://cyllene.uwa.edu.au/~dpannell/dp0901.htm.

Randall A (1987) *Resource Economics: An Economic Approach to Natural Resource and Environmental Policy.* 2nd edn. Wiley: New York.

Ridley AM and Pannell DJ (2005) The role of plants and plant-based R&D in managing dryland salinity in Australia. *Australian Journal of Experimental Agriculture* **45**(11), 1341–1355.

Roberts AM and Pannell DJ (2009) Piloting a systematic framework for public investment in regional natural resource management: dryland salinity in Australia. *Land Use Policy* **26**(4), 1001–1010.

Stoneham G, Chaudhri V, Ha A and Strappazzon L (2003) Auctions for conservation contracts: and empirical examination of Victoria's BushTender trial. *Australian Journal of Agricultural and Resource Economics* **47**(4), 477–500.

Index

www.ingramcontent.com/pod-product-compliance
Lightning Source LLC
Chambersburg PA
CBHW052141170526
45159CB00017B/3131